Data Analysis

Data Analysis

A gentle introduction for future data scientists

Graham Upton

Former Professor of Applied Statistics, University of Essex

and

Dan Brawn

Lecturer, Department of Mathematical Sciences, University of Essex

OXFORD
UNIVERSITY PRESS

Great Clarendon Street, Oxford, OX2 6DP,
United Kingdom

Oxford University Press is a department of the University of Oxford.
It furthers the University's objective of excellence in research, scholarship,
and education by publishing worldwide. Oxford is a registered trade mark of
Oxford University Press in the UK and in certain other countries

Published in the United States of America by Oxford University Press
198 Madison Avenue, New York, NY 10016, United States of America

British Library Cataloguing in Publication Data
Data available

Library of Congress Control Number: 2023931425

ISBN 978–0–19–288577–7
ISBN 978–0–19–288578–4 (pbk.)

DOI: 10.1093/oso/9780192885777.001.0001

Printed and bound by
CPI Group (UK) Ltd, Croydon, CR0 4YY

Contents

Preface   ix

1. First steps   1
1.1 Types of data   1
1.2 Sample and population   2
    1.2.1 Observations and random variables   3
    1.2.2 Sampling variation   3
1.3 Methods for sampling a population   4
    1.3.1 The simple random sample   5
    1.3.2 Cluster sampling   5
    1.3.3 Stratified sampling   6
    1.3.4 Systematic sampling   7
1.4 Oversampling and the use of weights   7

2. Summarizing data   9
2.1 Measures of location   9
    2.1.1 The mode   9
    2.1.2 The mean   10
    2.1.3 The trimmed mean   11
    2.1.4 The Winsorized mean   11
    2.1.5 The median   12
2.2 Measures of spread   12
    2.2.1 The range   12
    2.2.2 The interquartile range   12
2.3 Boxplot   13
2.4 Histograms   14
2.5 Cumulative frequency diagrams   16
2.6 Step diagrams   17
2.7 The variance and standard deviation   18
2.8 Symmetric and skewed data   20

3. Probability   21
3.1 Probability   21
3.2 The rules of probability   24
3.3 Conditional probability and independence   25
3.4 The total probability theorem   28
3.5 Bayes' theorem   30

4. Probability distributions — 33
4.1 Notation — 33
4.2 Mean and variance of a probability distribution — 33
4.3 The relation between sample and population — 35
4.4 Combining means and variances — 37
4.5 Discrete uniform distribution — 39
4.6 Probability density function — 39
4.7 The continuous uniform distribution — 41

5. Estimation and confidence — 43
5.1 Point estimates — 43
    5.1.1 Maximum likelihood estimation (mle) — 43
5.2 Confidence intervals — 43
5.3 Confidence interval for the population mean — 44
    5.3.1 The normal distribution — 44
    5.3.2 The Central Limit Theorem — 45
    5.3.3 Construction of the confidence interval — 48
5.4 Confidence interval for a proportion — 48
    5.4.1 The binomial distribution — 49
    5.4.2 Confidence interval for a proportion (large sample case) — 50
    5.4.3 Confidence interval for a proportion (small sample) — 50
5.5 Confidence bounds for other summary statistics — 50
    5.5.1 The bootstrap — 50
5.6 Some other probability distributions — 52
    5.6.1 The Poisson and exponential distributions — 52
    5.6.2 The Weibull distribution — 55
    5.6.3 The chi-squared (χ^2) distribution — 56

6. Models, p-values, and hypotheses — 59
6.1 Models — 59
6.2 p-values and the null hypothesis — 60
    6.2.1 Two-sided or one-sided? — 60
    6.2.2 Interpreting p-values — 61
    6.2.3 Comparing p-values — 62
    6.2.4 Link with confidence interval — 63
6.3 p-values when comparing two samples — 63
    6.3.1 Do the two samples come from the same population? — 63
    6.3.2 Do the two populations have the same mean? — 65

7. Comparing proportions — 67
7.1 The 2×2 table — 67
7.2 Some terminology — 69
    7.2.1 Odds, odds ratios, and independence — 70

7.2.2 Relative risk   70
7.2.3 Sensitivity, specificity, and related quantities   71
7.3 The $R \times C$ table   72
7.3.1 Residuals   73
7.3.2 Partitioning   74

8. Relations between two continuous variables   77

8.1 Scatter diagrams   78
8.2 Correlation   79
8.2.1 Testing for independence   81
8.3 The equation of a line   84
8.4 The method of least squares   84
8.5 A random dependent variable, Y 87
8.5.1 Estimation of σ^2 88
8.5.2 Confidence interval for the regression line   88
8.5.3 Prediction interval for future values   88
8.6 Departures from linearity   89
8.6.1 Transformations   89
8.6.2 Extrapolation   90
8.6.3 Outliers   91
8.7 Distinguishing x and Y 93
8.8 Why 'regression'?   93

9. Several explanatory variables   97

9.1 *AIC* and related measures   98
9.2 Multiple regression   99
9.2.1 Two variables   99
9.2.2 Collinearity   101
9.2.3 Using a dummy variable   102
9.2.4 The use of multiple dummy variables   104
9.2.5 Model selection   107
9.2.6 Interactions   107
9.2.7 Residuals   109
9.3 Cross-validation   110
9.3.1 k-fold cross-validation   110
9.3.2 Leave-one-out cross-validation (LOOCV)   112
9.4 Reconciling bias and variability   112
9.5 Shrinkage   113
9.5.1 Standardization   114
9.6 Generalized linear models (GLMs)   115
9.6.1 Logistic regression   116
9.6.2 Loglinear models   118

10. Classification   121

10.1 Naive Bayes classification   121
10.2 Classification using logistic regression   124

10.3 Classification trees   125
10.4 The random forest classifier   127
10.5 *k*-nearest neighbours (*k*NN)   128
10.6 Support-vector machines   131
10.7 Ensemble approaches   133
10.8 Combining variables   134

11. Last words   **135**
Further reading   138

Index   139

Preface

This book aims to provide the would-be data scientist with a working idea of the most frequently used tools of data analysis. Our aim has been to introduce approaches to data analysis with a minimum of equations.

We envisage three general classes of readers: the complete novice, those with existing statistical knowledge, and those currently employed as data scientists or using data science who wish to widen their repertoire and learn something of the underlying methodology.

The entire book is relevant for the novice. The earlier chapters may be most relevant for the practitioner, since they provide the background to the methods used in later chapters, while for statisticians new to data science it will be the final chapters that are of most use.

As a data scientist you will be using the computer to perform the data analysis. **Any programming language should be able to carry out the analyses that we describe.** We used R (because it is free); our code is available as an accompaniment to the book.

We hope you find this book useful.

1
First steps

This book assumes that you, the readers, are keen to become Data Scientists, but may have limited knowledge of Statistics or Mathematics. Indeed, you may have limited knowledge of what, exactly, is this new subject called Data Science.

According to *Wikipedia*,

> 'Data are individual facts, statistics, or items of information, often numeric, that are collected through observation.'

So the first thing to learn about the word 'Data' is that it is a plural.[1] The same source gives us

> 'Science is a systematic enterprise that builds and organizes knowledge in the form of testable explanations and predictions about the world.'

So we see that **the job of the Data Scientist is to form testable explanations of a mass of information.**

We will begin by looking at the various types of data and how those data may have been obtained.

1.1 Types of data

There are three common types: qualitative, discrete, and continuous.

Qualitative data (also referred to as **categorical data**) consist of descriptions using *names*. For example:

'Male' or 'Female'
'Oak', 'Ash', or 'Birch'

---

[1] It is the plural of 'Datum', which is rarely used

Data Analysis: A Gentle Introduction for Future Data Scientists. Graham Upton and Dan Brawn, Oxford University Press.
© Graham Upton and Dan Brawn (2023). DOI: 10.1093/oso/9780192885777.003.0001

In the examples above, the alternatives are just names and the variables might be referred to as **nominal variables**. In other cases the categories may have a clear ordering:

'Poor', 'Adequate', or 'Abundant'
'Small', 'Medium', or 'Large'
'Cottage', 'House', or 'Mansion'

In these cases the variables are called **ordinal** variables.

Discrete data consist of numerical values in cases where we can make a list of the possible values. Often the list is very short:

1, 2, 3, 4, 5, and 6.

Sometimes the list will be infinitely long, as for example the list

0, 0.5, 1, 1.5, 2, 2.5, 3.0, 3.5, 4.0, ...

Continuous data consist of numerical values in cases where it is not possible to make a list of the outcomes. Examples are measurements of physical quantities such as weight, height, and time.

- *The distinction between discrete and continuous data is often blurred by the limitations of our measuring instruments. For example, we may record our heights to the nearest centimetre, in which case observations of a continuous quantity are being recorded using discrete values.*

1.2 Sample and population

Here the term 'population' refers to a complete collection of anything! Could be pieces of paper, pebbles, tigers, sentence lengths.... Absolutely anything, not just people. In the next paragraph, the population is 'the heights of every oak tree in the country'.

Suppose we want to know what proportion of oak trees are greater than 10 metres in height. We cannot examine every oak tree in the country (the population). Instead we take a sample consisting of a relatively small number of oak trees selected randomly from the population of oak trees. We find p, the proportion of trees in the sample that are greater than 10 metres in height. We hope that our sample is representative of the population (we discuss how to efficiently randomly sample later in the chapter). We therefore take p as our estimate of the proportion of oak trees in the population that are greater than 10 metres in height.

1.2.1 Observations and random variables

The terms **observation** or **observed value** are used for the individual values within our sample. Different samples will usually give rise to different sets of observations. Imagine all the samples of oak trees that might have been observed. Consider the first observation in the sample actually taken. That tree could have been any one of the oak trees in the population (which consists of every tree that might be observed). If the sample were taken at random, then the height recorded could have been any of the heights of the oak trees in the population. These heights vary and our choice has been made at random: we have an observation on a **random variable**.

Some other examples are:

Random variable	Range of values	Value observed
Weight of a person	Perhaps 45 kg to 180 kg	80 kg
Speed of a car in the UK	0 to 70 mph (legally!)	69 mph
Number of letters in a post box	Perhaps 0 to 100	23
Colour of a postage stamp	All possible colours	Tyrian plum

1.2.2 Sampling variation

Suppose that we have carefully taken a random sample from the population of oak trees. Using the heights of the trees in our sample, we have happily obtained an estimate, p, of the proportion greater than 10 metres in height. That's fine, but if we take another sample, with equal care, we would not expect to get *exactly* the same proportion (though we would expect it to be similar). If we take several samples then we can get an idea of a range of values within which the true population proportion lies. We will see in later chapters that we can often deduce that range from a single sample.

Example 1.1

Suppose that a large population contains equal proportions of 1, 2, 3, 4, and 5. A random sample of just four observations is taken from the population. The observations in this sample are 1, 5, 1, and 1. These have an average of 2. The next three samples are (2, 4, 2, 2), (1, 4, 1, 5), and (4, 2, 2, 3) with averages 2.5, 2.75, and 2.75, respectively. A further 96 samples, each containing four observations are randomly generated. The 100 averages are illustrated using a bar chart in Figure 1.1.

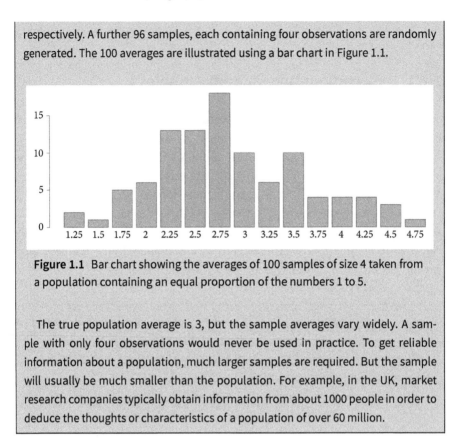

Figure 1.1 Bar chart showing the averages of 100 samples of size 4 taken from a population containing an equal proportion of the numbers 1 to 5.

The true population average is 3, but the sample averages vary widely. A sample with only four observations would never be used in practice. To get reliable information about a population, much larger samples are required. But the sample will usually be much smaller than the population. For example, in the UK, market research companies typically obtain information from about 1000 people in order to deduce the thoughts or characteristics of a population of over 60 million.

- *As we saw in the example, different samples from the same population are likely to have different means. Each mean is an estimate of the population mean (the average of all the values in the population).*
- *The larger the sample size, the nearer (on average) will the sample mean be to the unknown population mean that it aims to estimate.*

1.3 Methods for sampling a population

Although the data analyst is unlikely to be collecting the data, it is always important to understand how the data were obtained. If the data consist of information on the entire population of interest, then this is straightforward. However, if the data represent information on a sample from the population, then it is important to know how the sample was obtained. Some common methods are briefly described in this section.

1.3.1 The simple random sample

Most sampling methods endeavour to give every member of the population a known probability of being included in the sample. If each member of the sample is selected by the equivalent of drawing lots, then the sample selected is described as being a **simple random sample**.

One procedure for drawing lots is the following:

1. Make a list of all N members of the population.
2. Assign each member of the population a different number.
3. For each member of the population place a correspondingly numbered ball in a bag.
4. Draw n balls from the bag, without replacement. The balls should be chosen at random.
5. The numbers on the balls identify the chosen members of the population.

An automated version would use the computer to simulate the drawing of the balls from the bag.

The principal difficulty with the procedure is the first step: the creation of a list of all N members of the population. This list is known as the **sampling frame**. In many cases there will be no such central list. For example, suppose it was desired to test the effect of a new cattle feed on a random sample of British cows. Each individual farm may have a list of its own cows (Daisy, Buttercup, ...), but the goverment keeps no central list.

Indeed, for the country as a whole there is not even a 100% accurate list of people (because of births, deaths, immigration, and emigration).

But, because of the straightforward nature of the simple random sample, it will be the first choice when it is practicable.

1.3.2 Cluster sampling

Even if there were a 100% accurate list of the British population, simple random sampling would almost certainly not be performed because of expense. It is easy to imagine the groans emitted by the pollsters on drawing two balls from the bag corresponding to inhabitants of Land's End and the Shetland Isles. The intrepid interviewer would be a much travelled individual!

To avoid this problem, populations that are geographically scattered are usually divided into conveniently sized regions. A possible procedure is then

1. Choose a region at random.
2. Choose individuals at random from that region.

The consequences of this procedure are that instead of a random scatter of selected individuals there are scattered **clusters** of individuals. The selection probabilities for the various regions are not equal but would be adjusted to be in proportion to the numbers of individuals that the regions contain. If there are r regions, with the ith region containing N_i individuals, then the chance that it is selected would be chosen to be N_i/N, where $N = N_1 + N_2 + \cdots + N_r$.

The size of the chosen region is usually sufficiently small that a single interviewer can perform all the interviews in that region without incurring huge travel costs. In practice, because of the sparse population and the difficulties of travel in the highlands and islands of Scotland, most studies of the British population are confined to the region south of the Caledonian Canal.

1.3.3 Stratified sampling

Most populations contain identifiable **strata**, which are non-overlapping subsets of the population. For example, for human populations, useful strata might be 'males' and 'females', or 'Receiving education', 'Working', and 'Retired', or combinations such as 'Retired female'. From census data we might know the proportions of the population falling into these different categories. With stratified sampling, we ensure that these proportions are reproduced by the sample. Suppose, for example, that the age distribution of the adult population in a particular district is as given in the table below.

Aged under 40	Aged between 40 and 60	Aged over 60
38%	40%	22%

A simple random sample of 200 adults would be unlikely to *exactly* reproduce these figures. If we were very unfortunate, over half the individuals in the sample might be aged under 40. If the sample were concerned with people's taste in music then, by chance, the simple random sample might provide a misleading view of the population.

A **stratified sample** is made up of separate simple random samples for each of the strata. In the present case, we would choose a simple random sample of 76 (38% of 200) adults aged under 40, a separate simple random sample of 80 adults aged between 40 and 60, and a separate simple random sample of 44 adults aged over 60.

Stratified samples exactly reproduce the characteristics of the strata and this almost always increases the accuracy of subsequent estimates of population parameters. Their slight disadvantage is that they are a little more difficult to organize.

1.3.4 Systematic sampling

Suppose that we have a list of the N members of a population. We wish to draw a sample of size n from that population. Let k be an integer close to N/n. Systematic sampling proceeds as follows:

1. Choose one member of the population at random. This is the first member of the population.
2. Choose every kth individual thereafter, returning to the beginning of the list when the end of the list is reached.

For example, suppose we wish to choose six individuals from a list of 250. A convenient value for k might be 40. Suppose that the first individual selected is number 138. The remainder would be numbers 178, 218, 8, 48, and 88.

If the population list has been ordered by some relevant characteristic, then this procedure produces a spread of values for the characteristic—a type of informal stratification.

1.4 Oversampling and the use of weights

Using either strata or clusters, the sampler is dividing a population into smaller sections. Often both strata and clusters are required, with the potential result that some subsamples would include only a few individuals. Small samples are inadvisable since, by chance, they may give a false idea of the subpopulations that they are supposed represent.

The solution is oversampling. The sample size for the rare subpopulation is inflated by some multiple M, so as to give a reliable view of that subpopulation. Subsequent calculations will take account of this oversampling so as to fairly represent the entire population.

Providing the sampling method is unbiased, large samples will be more accurate than small samples. However, a small unbiased sample would usually be preferred to a larger biased sample.

2
Summarizing data

Now we have a computer full of data, we have to decide what to do with it! How should we report it? How should we present it to those who need to know? In this chapter we present possible answers to these questions. We apologize in advance for the number of specialized terms and their definitions.

2.1 Measures of location

2.1.1 The mode

The **mode** of a set of discrete data is the single outcome that occurs most frequently. This is the simplest of the measures of location, but it is of limited use. If there are two such outcomes that occur with equal frequency then there is no unique mode and the data would be described as being **bimodal**; if there were three or more such outcomes then the data would be called **multimodal**.

When measured with sufficient accuracy all observations on a continuous variable will be different: even if John and Jim both *claim* to be 1.8 metres tall, we can be certain that their heights will not be *exactly* the same. However, if we plot a histogram (Section 2.4) of men's heights, we will usually find that it has a peak in the middle: the class corresponding to this peak is called the **modal class**.

Example 2.1

A random sample of 50 place names was chosen from a gazetteer of place names in north-west Wales. The numbers of letters in these place names are summarized in the following table:

Length of place name	4	5	6	7	8	9	10	11	12	13	14	15	16	17
Frequency of occurrence	3	5	5	5	6	8	9	3	1	1	2	1	0	1

Data Analysis: A Gentle Introduction for Future Data Scientists. Graham Upton and Dan Brawn, Oxford University Press.
© Graham Upton and Dan Brawn (2023). DOI: 10.1093/oso/9780192885777.003.0002

In this sample, the most frequent length of place name is 10 (often names starting Llan), which is therefore the mode. According to Wikipedia there are over 600 places in Wales that start with 'Llan' which refers to the region around a church.

2.1.2 The mean

This measure of location is simply the **average** value: the sum of the observed values divided by the number of values in the sample. Unlike the mode, the mean will usually not be equal to any of the observed values.

Suppose that the data set consists of the n values, $x_1, x_2, \ldots, x_n$. The sample mean, denoted by $\bar{x}$, is given by:

$$\bar{x} = \frac{1}{n}(x_1 + x_2 + \cdots + x_n) = \frac{1}{n}\sum_{i=1}^{n} x_i. \tag{2.1}$$

The convenient capital sigma notation,

$$\sum_{i=1}^{n}$$

states that summation is needed over the index i from 1 to n.

If the data are summarized in the form of a frequency distribution, with m distinct values, and with the value x_j occurring f_j times, then, since $\sum_{j=1}^{m} f_j = n$,

$$\bar{x} = \frac{1}{n}\sum_{j=1}^{m} f_j x_j. \tag{2.2}$$

This is just another way of writing Equation 2.1.[1] The formula would also be used for grouped data, with x_j being the mid-point of the jth group.

Example 2.1 (cont.)

The sum of the lengths of the 50 Welsh place names is $(3 \times 4) + (5 \times 5) + \cdots + (1 \times 17) = 430$ so the mean length is $430/50 = 8.6$ letters.

---

[1] All that is happening is that $y + y + y$, say, is being calculated as $3 \times y$.

2.1.3 The trimmed mean

Sometimes the data contain a number of **outliers** (values that are not typical of the bulk of the data). For example, if a multi-millionaire takes up residence in a small village, then the average income for the village inhabitants will be greatly inflated and the resulting value would not be useful. A trimmed mean avoids this problem by discarding a specified proportion of both the largest and the smallest values.

Example 2.2

The numbers of words in the 18 sentences of Chapter 1 of *A Tale of Two Cities* by Charles Dickens are as follows:

118, 39, 27, 13, 49, 35, 51, 29, 68, 54, 58, 42, 16, 221, 80, 25, 41, 33.

The famous first sentence ('It was the best of times, it was the worst of times, …') is unusually long, but the fourteenth sentence is even longer. The rest of the sentences have more usual lengths. The mean is 55.5, which is greater than 13 of the 18 values.

A 20% trimmed mean would ignore the bottom 10% and the top 10% of values. Taking 10% as meaning just one of the 18 values, the trimmed mean is in this case the average of the central 16 values: 47.8. This is much more representative, being larger than 10 values and less than 8.

2.1.4 The Winsorized mean

In this alternative to the trimmed mean, the mean is calculated after replacing the extreme values by their less extreme neighbours. When data are collected simultaneously on several variables (this is called **multivariate data**) and one variable appears to contain unusual values, Winsorizing could be preferable to trimming because it will retain the correct number of 'observed' values. However, the Winsorizing must be reported as part of the data analysis.

Example 2.2 (cont.)

A 20% Winsorized mean for the previous data would use the values

16, 16, 25, 27, 29, 33, 35, 39, 41, 42, 49, 51, 54, 58, 68, 80, 118, 118.

to report a value of 49.9.

2.1.5 The median

The word 'median' is just a fancy name for 'middle'. It is a useful alternative (because it is easier to explain) to the trimmed mean (or the Winsorized mean) when there are outliers present.

With n observed values arranged in order of size, the median is calculated as follows. If n is odd and equal to $(2k+1)$, say, then the median is the $(k+1)$th ordered value. If n is even and equal to $2k$, say, then the median is the average of the kth and the $(k+1)$th ordered values.

Example 2.2 (cont.)

Re-ordering the 18 sentence lengths gives

13, 16, 25, 27, 29, 33, 35, 39, 41, 42, 49, 51, 54, 58, 68, 80, 118, 221.

The median length is 41.5 words.

2.2 Measures of spread

2.2.1 The range

This usually refers to a pair of values: the smallest value and the largest value.

Example 2.2 (cont.)

The range of sentence lengths is from 13 to 221, a span of 221–13 = 208 words.

2.2.2 The interquartile range

The median (Section 2.1.5) is the value that subdivides ordered data into two halves. The **quartiles** go one step further by using the same rules to divide each half into halves (to form quarters).

Denoting the **lower quartile** and **upper quartile** by Q_1 and Q_2, respectively, the inter-quartile range is reported either as the pair of values (Q_1, Q_2), or as the difference $Q_2 - Q_1$.

Quartiles are a special case of **quantiles**. Quantiles subdivide the ordered observations in a sample into equal-sized chunks. Another example of quantiles is provided by the **percentiles** which divide the ordered data into 100 sections. Quartiles are useful for the construction of the boxplots introduced in the next section.

Example 2.2 (cont.)

Using R, the lower and upper quartiles were reported to be 30 and 57, respectively. Different computer packages may report slightly different values.

2.3 Boxplot

This is a diagram that summarizes the variability in the data. It is also known as a **box-whisker diagram**. The diagram consists of a central box with a 'whisker' at each end.

The ends of the central box are the lower and upper quartiles, with either a central line or a notch indicating the value of the median. The simplest form of the diagram has two lines (the whiskers); one joining the lowest value to the lower quartile, and the other joining the highest value to the upper quartile.

A refined version has the length of the whiskers limited to some specified multiple of the inter-quartile range. In this version any more extreme values are shown individually.

Box-whisker diagrams provide a particularly convenient way of comparing two or more sets of values. In this case, a further refinement may vary the widths of the boxes to be proportional to the numbers of observations in each set.

Example 2.2 (cont.)

An interesting contrast to the lengths of sentences in the Dickens work is provided by the lengths of the seventeen sentences in Chapter 1 of Jeffrey Archer's *Not a Penny More, Not a Penny Less*:

$$8, 10, 15, 13, 32, 25, 14, 16, 32, 25, 5, 34, 36, 19, 20, 37, 19.$$

When the whisker lengths are limited to a maximum of 1.5 times the inter-quartile range, the result is Figure 2.1.

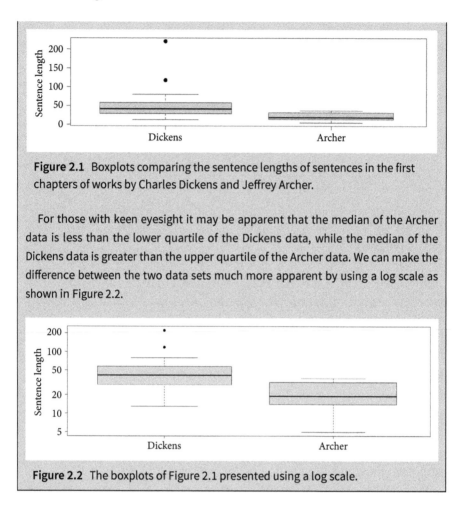

Figure 2.1 Boxplots comparing the sentence lengths of sentences in the first chapters of works by Charles Dickens and Jeffrey Archer.

For those with keen eyesight it may be apparent that the median of the Archer data is less than the lower quartile of the Dickens data, while the median of the Dickens data is greater than the upper quartile of the Archer data. We can make the difference between the two data sets much more apparent by using a log scale as shown in Figure 2.2.

Figure 2.2 The boxplots of Figure 2.1 presented using a log scale.

2.4 Histograms

Bar charts are not appropriate for **continuous data** (such as measurements of weight, time, etc.). A **histogram** is a diagram that also uses rectangle areas to represent frequency. It differs from the bar chart in that the rectangles may have differing widths, but the key feature is that, for each rectangle,

Area is proportional to **class frequency**.

Some computer packages attempt to make histograms three-dimensional. Avoid these if you can, since the effect is likely to be misleading.

Example 2.3

As glaciers retreat they leave behind rocks known as 'erratics' because they are of a different type to the rocks normally found in the area. In Scotland many of these erratics have been used by farmers in the walls of their fields. One measure of the size of these erratics is the cross-sectional area visible on the outside of a wall. The areas of 30 erratics (in cm^2) are given below:

> 216, 420, 240, 100, 247, 128, 540, 594, 160, 286, 216, 448, 380, 509, 90, 156, 135, 225, 304, 144, 152, 143, 135, 266, 286, 154, 154, 386, 378, 160

Area is a continuous variable, so a histogram is appropriate. Figure 2.3 uses six intervals, each spanning 100 cm^2.

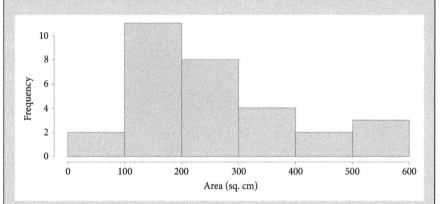

Figure 2.3 Histogram of the data on the cross-sections of erratics, using equal-width ranges.

The interpretation of the y-axis is that there are, for example, eight observations with values between 200 cm^2 and 300 cm^2. Of course the figure is not lying, but it is, perhaps, giving a false impression of the distribution of the areas. A better idea is provided by collecting the data into decades (0–99, 100–109, etc.). The result is Table 2.1.

Table 2.1 Summary of the cross-section data for the erratics. Here '9–' means an observation in the range 90–99.

Decade	9–	10–	12–	13–	14–	15–	16–	21–	22–	24–
Count	1	1	1	2	2	4	2	2	1	2
Decade	26–	28–	30–	37–	38–	42–	44–	50–	54–	59–
Count	1	2	1	1	2	1	1	1	1	1

The table shows that areas around 150 cm^2 are particularly frequent. To bring this out in the histogram requires unequal intervals. For Figure 2.4 we used ranges with the following end-points: (0, 125, 150, 200, 250, 400, 600).

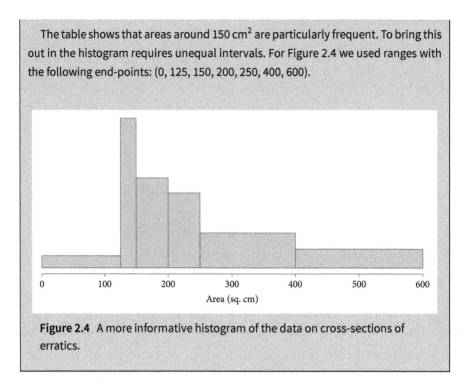

Figure 2.4   A more informative histogram of the data on cross-sections of erratics.

It is always a good idea to experiment with varying diagrams.

2.5 Cumulative frequency diagrams

These are an alternative form of diagram that provide answers to questions such as 'What proportion of the data have values less than x?'. In such a diagram, cumulative frequency on the 'y-axis' is plotted against observed value on the 'x-axis'. The result is a graph in which, as the x-coordinate increases, the y-coordinate cannot decrease.

With grouped data, such as in the following example, the first step is to produce a table of cumulative frequencies. These are then plotted against the corresponding upper class boundaries. The successive points may be connected either by straight-line joins (in which case the diagram is called a **cumulative frequency polygon**) or by a curve (in which case the diagram is called an **ogive**).

Example 2.4

In studying bird migration, a standard technique is to put coloured rings around the legs of the young birds at their breeding colony. Table 2.2, which refers to recoveries of razorbills, summarizes the distances (measured in hundreds of miles) between the recovery point and the breeding colony.

Table 2.2 Recovery distances of ringed razorbills.

Distance (miles) (x)	Freq.	Cum. freq.	Distance (miles) (x)	Freq.	Cum. freq.
$x < 100$	2	2	$700 \leq x < 800$	2	30
$100 \leq x < 200$	2	4	$800 \leq x < 900$	2	32
$200 \leq x < 300$	4	8	$900 \leq x < 1000$	0	32
$300 \leq x < 400$	3	11	$1000 \leq x < 1500$	2	34
$400 \leq x < 500$	5	16	$1500 \leq x < 2000$	0	34
$500 \leq x < 600$	7	23	$2000 \leq x < 2500$	2	36
$600 \leq x < 700$	5	28			

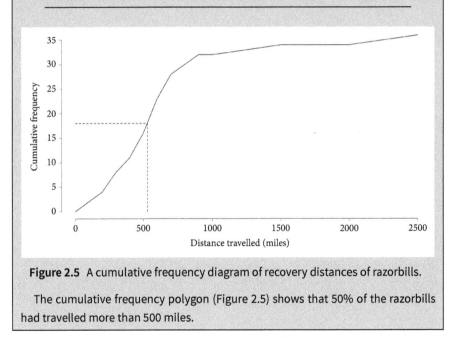

Figure 2.5 A cumulative frequency diagram of recovery distances of razorbills.

The cumulative frequency polygon (Figure 2.5) shows that 50% of the razorbills had travelled more than 500 miles.

2.6 Step diagrams

A cumulative frequency diagram for ungrouped data is sometimes referred to as a **step polygon** or **step diagram** because of its appearance.

Example 2.5

In a compilation of Sherlock Holmes stories, the 13 stories that comprise *The Return of Sherlock Holmes* have the following numbers of pages:

13.7, 15.5, 16.4, 12.8, 20.8, 13.7, 11.2, 13.7, 11.7, 15.0, 14.1, 14.8, 17.1.

The lengths are given to the nearest tenth of a page. Treating the values as being exact, we use them as the boundaries in a cumulative frequency table. We first need to order the values:

11.2, 11.7, 12.8, 13.7, 13.7, 13.7, 14.1, 14.8, 15.0, 15.5, 16.4, 17.1, 20.8.

The resulting diagram is Figure 2.6.

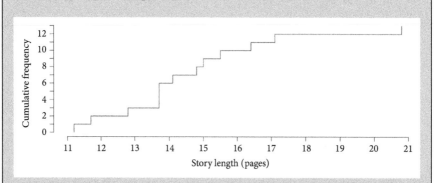

Figure 2.6 A step diagram illustrating the lengths of stories in *The Return of Sherlock Holmes*.

Notice that the cumulative frequencies 'jump' at each of the observed values. It is this that gives rise to the vertical strokes in the diagram. The horizontal strokes represent the ranges between the ordered values.

2.7 The variance and standard deviation

Denoting a sample of n values by $x_1, x_2, \ldots, x_n$, and their average (the sample mean) by $\bar{x}$, the variability is captured by the differences of the values from their mean: $(x_1 - \bar{x}), (x_2 - \bar{x}), \ldots$.

To avoid positive differences cancelling out negative differences, we work with the squared values: $(x_1 - \bar{x})^2, (x_2 - \bar{x})^2, \ldots$.[2] To get a measure of the overall variability, we work with their sum:

$$(x_1 - \bar{x})^2 + (x_2 - \bar{x})^2 + \cdots + (x_n - \bar{x})^2 = \sum_{i=1}^{n}(x_i - \bar{x})^2.$$

The magnitude of this sum is affected by the number of terms in the summation. To get a proper idea of the average variability we need to take the size of n into account. This we do by dividing by $(n - 1)$ to obtain the quantity s^2 given by[3]:

$$s^2 = \frac{1}{n - 1}\sum_{i=1}^{n}(x_i - \bar{x})^2. \tag{2.3}$$

This has a minimum value of zero when all the values are equal.

The units of variance are squared units. To obtain an idea of the variability in terms of the units of the observations, all that is needed is to take the square root. This quantity, denoted by s, is called the **standard deviation** (or **sample standard deviation**).

As a general rule (as a consequence of the Central Limit Theorem; see Section 5.3.2) you can expect most data to lie in the range $(\bar{x} - 2s, \bar{x} + 2s)$. Observations less than $\bar{x} - 3s$, or greater than $\bar{x} + 3s$ will be very unusual. If any occur then it would be wise to check that they have been recorded correctly. Even with computers, transcription errors can occur.

Example 2.5 (cont.)

To investigate the general rule given above, for one final time we look at the sentence lengths. For the Archer data, $(\bar{x} - 2s, \bar{x} + 2s)$ corresponds to the interval (1, 42) which does indeed contain all the Archer sentence lengths. For the Dickens data, because of two extraordinarily long sentences, the interval is the bizarre (-41, 150). The lower value is preposterous and alerts us to the unusual nature of the data. Only the longest sentence falls outside this range.

---

[2] We cannot work with the sum of the differences as this is zero. We could work with the sum of the absolute differences, but that leads to intractable mathematics!

[3] The reason for using $(n - 1)$ rather than n is that we want s^2, which is a sample quantity, to be an unbiased estimate of the same quantity in the population being sampled.

> *There is no need to report numbers accurately when they are only being used as a*
> *guide. In general, two significant figures will suffice.*

2.8 Symmetric and skewed data

If a population is approximately **symmetric** then a reasonable sized sample
will have mean and median having similar values. Typically their values will
also be close to that of the mode of the population (if there is one!).

A population that is not symmetric is said to be **skewed**. A distribution
with a long 'tail' of high values (like the Dickens data) is said to be **positively skewed**, in which case the mean is usually greater than the mode or
the median. If there is a long tail of low values then the mean is likely to be
the lowest of the three location measures and the distribution is said to be
negatively skewed.

3
Probability

3.1 Probability

Although the data analyst is unlikely to need to perform complex probability calcula-
tions, a general understanding of probability is certainly required.

Suppose we have an ordinary six-sided die. When we roll it, it will show
one of six faces:

Suppose we roll the die six times. Using numbers, as they are easier to read,
we get: 1, 5, 1, 1, 2, and 4. Have we got one of each face? Certainly not, by
chance (actually using a random number generator) we find that half of the
numbers are 1s and there are no 3s or 6s. Since we are using the computer, we
can find out what happens as we 'roll the die' many more times (see Table 3.1).

If the die is fair, then, on average, and in the very long run, each face will
appear on exactly 1/6 of occasions. After 60,000 repeats we are very close to
the limiting value.[1]

- It is often helpful to imagine probability as being the limiting value of the proportion
  approached over a long series of repeats of identical experiments.
- Sometimes, the probability assigned to an event is no more than an informed judgement: for example, the probability of an individual being struck by a meteorite.

Table 3.1 The results of 'rolling a six-sided die' (using a computer).

Number of observations	Proportion showing a particular face					
	1	2	3	4	5	6
6	0.500	0.167	0.000	0.167	0.167	0.000
60	0.133	0.283	0.100	0.183	0.167	0.133
600	0.162	0.195	0.160	0.153	0.163	0.167
6000	0.168	0.174	0.166	0.172	0.165	0.156
60,000	0.169	0.166	0.167	0.165	0.168	0.165

[1] The value that we would get after an infinite number of repeats (supposing we were still alive!).

Data Analysis: A Gentle Introduction for Future Data Scientists. Graham Upton and Dan Brawn, Oxford University Press.
© Graham Upton and Dan Brawn (2023). DOI: 10.1093/oso/9780192885777.003.0003

We start with some formal definitions and notation, using as an example the result of a single throw of a fair six-sided die:

- The **sample space**, S, is the set of all possible outcomes of the situation under investigation. At least one of the possible outcomes must occur.
  In the example the six possible outcomes are 1, 2, 3, 4, 5, 6.
- An **event**, E, is a possible outcome, or group of outcomes, of special interest.
  For example, obtaining a 6.

Probability ranges from 0 to 1:

- An event or value that cannot occur has probability 0.
  Obtaining a 7.
- An event or value that is certain to occur has probability 1.
  Obtaining a number in the range 1 to 6 inclusive.
- All other events have probabilities between 0 and 1.
  For example, obtaining a 6.
- An event or value that has an equal chance of occurring or not occurring has probability 0.5.
  Obtaining an odd number.

Diagrams that help with visualising the relation between events are called **Venn diagrams.**[2] Figure 3.1 is a Venn diagram illustrating the sample space for the die-rolling example. The sample space is subdivided into six equal-sized parts corresponding to the six possible outcomes. The highlighted area is one-sixth of the whole: $P(\text{die shows a } 5) = 1/6$.

To handle probabilities involving more than one event we need some more notation:

- $A \cup B$ A **union** B At least one of events A and B occurs.
- $A \cap B$ A **intersection** B Events A and B both occur.

Figure 3.2 uses Venn diagrams to illustrate the union and intersection ideas. Suppose that the probability that an event X occurs is denoted by $P(X)$. Then, since the intersection, $A \cap B$, is part of both A and B, we can see that:

$$P(A) + P(B) = P(A \cup B) + P(A \cap B). \tag{3.1}$$

Rearranging, this means that:

$$P(A \cup B) = P(A) + P(B) - P(A \cap B). \tag{3.2}$$

---

[2] John Venn (1834–1923) was a Cambridge lecturer whose major work, *The Logic of Chance*, was published in 1866. It was in this book that he introduced the diagrams that now bear his name.

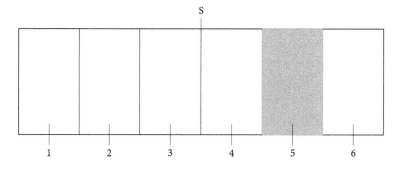

Figure 3.1 Venn diagram illustrating the sample space S and its subdivision into six equi-probable parts including the event 'obtaining a 5'.

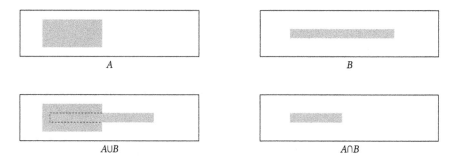

Figure 3.2 Venn diagrams illustrating events A, B, their union $A \cup B$, and their intersection $A \cap B$.

Example 3.1

Continuing with the fair six-sided die, define the events A and B as follows:

A : The die shows a multiple of 2.   2, 4, or 6

B : The die shows a multiple of 3.   3 or 6

Thus $P(A) = \frac{3}{6}$ and $P(B) = \frac{2}{6}$.

Combining the events:

$A \cup B$: The die shows a multiple of 2 or 3 (or both).   2, 3, 4, or 6

$A \cap B$: The die shows both a multiple of 2 and a multiple of 3.   6

Thus $P(A \cup B) = \frac{4}{6}$ and $P(A \cap B) = \frac{1}{6}$.

Alternatively, using Equation (3.2), $P(A \cup B) = \frac{3}{6} + \frac{2}{6} - \frac{1}{6} = \frac{4}{6}$.

3.2 The rules of probability

> **Addition rule**: *If events are mutually exclusive, then the probability that one or other occurs is the sum of the probabilities of the individual events.*

If events C and D are **mutually exclusive** then they cannot occur simultaneously, which implies that $P(C \cap D) = 0$ (see Figure 3.3). The rule is a consequence of substituting zero for the intersection term in Equation (3.2).

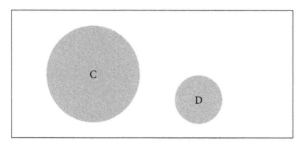

Figure 3.3 Events C and D have no intersection: they are mutually exclusive.

Example 3.2

We roll a fair six-sided die that has sides numbered 1 to 6. The die can only show one side at a time, so the probability of occurrence for the event 'the die shows either a 6 or a 1 on top' is P(die shows a 6) + P(die shows a 1)$= \frac{1}{6} + \frac{1}{6} = \frac{1}{3}$.

> **Multiplication rule**: *If events are mutually independent (i.e., the probability of an event occurring is the same, regardless of what other events occur), then the probability that all the events occur is the product of their separate probabilities.*

Example 3.3

Suppose we now roll a fair six-sided die twice. On the first roll we are equally likely to obtain any of 1 to 6, so that P(6)$= \frac{1}{6}$. The same is true for the second roll: P(6)$= \frac{1}{6}$. The outcomes of the rolls are independent of one another, so the probability that we obtain 6 on both rolls is $\frac{1}{6} \times \frac{1}{6} = \frac{1}{36}$. Figure 3.4 illustrates the situation.

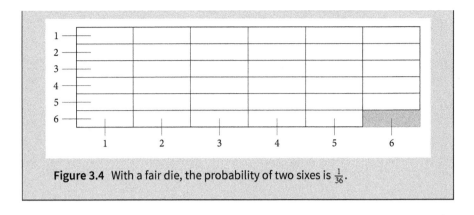

Figure 3.4 With a fair die, the probability of two sixes is $\frac{1}{36}$.

If $E_1, E_2, \ldots,$ are exclusive events in the sample space, S, and if these are the only events in S, then the sum of their probabilities is 1.

Example 3.4

Let E_i ($i = 1, 2, \ldots, 6$) be the event that side i is uppermost when a biased die is rolled. Because of the bias, we do not know the probability of any of the six events, but we do know that

$$P(E_1) + P(E_2) + \cdots + P(E_6) = 1.$$

3.3 Conditional probability and independence

The probability that we associate with the occurrence of an event is always likely to be influenced by any prior information that we have available. Suppose, for example, that I see a man lying motionless on the grass in a nearby park and am interested in the probability of the event 'the man is dead'. In the absence of other information a reasonable guess might be that the probability is one in a million. However, if I have just heard a shot ring out, and a suspicious-looking man with a smoking revolver is standing nearby, then the probability would be rather higher.

The probability that the event B occurs (or has occurred) given the information that the event A occurs (or has occurred) is written as $P(B|A)$.

Here $B|A$ is read as 'B **given** A' while P $(B|A)$ is described as a **conditional probability** since it refers to the probability that B occurs (or has occurred) *conditional* on the event that A occurs (or has occurred).

Example 3.5

Continuing with the example of a single throw of a fair six-sided die, define the events A and B as follows:

A: An odd number is obtained.

B: The number 3 is obtained.

If we do not know that A has occurred, then $P(B) = \frac{1}{6}$ as illustrated:

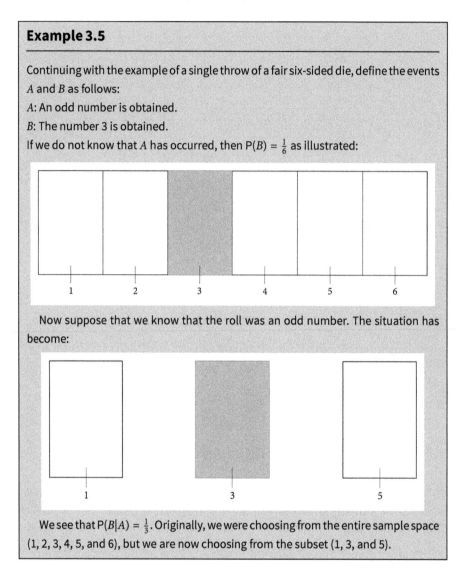

Now suppose that we know that the roll was an odd number. The situation has become:

We see that $P(B|A) = \frac{1}{3}$. Originally, we were choosing from the entire sample space (1, 2, 3, 4, 5, and 6), but we are now choosing from the subset (1, 3, and 5).

If it is known that event A occurs, then, as the previous example illustrated, the question is 'What proportion of the occasions when event A occurs, does event B also occur?' We define the answer to this question as the probability of event B conditional on event A, thus:

$$P(B|A) = \frac{P(A \cap B)}{P(A)}, \tag{3.3}$$

or, cross-multiplying,

$$P(A \cap B) = P(B|A) \times P(A). \tag{3.4}$$

Example 3.6

An electronic display is equally likely to show any of the digits $1, 2, \ldots, 9$. This time the event A is that an odd number is obtained, while the event B is that the number is prime.

P(A) that it shows an odd number (1, 3, 5, 7, or 9), is 5/9.
P(B), that it shows a prime number (2, 3, 5, or 7), is 4/9.
P($A \cap B$), that it shows an odd prime number (3, 5, or 7), is 3/9.

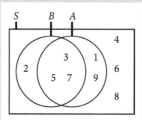

Figure 3.5 Venn diagram illustrating the prime number example.

Suppose that we are told that the number shown is odd. There are five equally likely odd numbers, so the conditional probability of a prime number, given this information, is 3/5. Alternatively using Equation (3.3):

$$P(B|A) = \frac{P(A \cap B)}{P(A)} = \frac{3/9}{5/9} = \frac{3}{5}.$$

Important links between independence and conditional probability:

- *If event B is independent of event A, then P($B|A$) = P(B).*
- *If P($B|A$) = P(B), then event B is independent of event A.*

The assumption of independence is central to many data analysis methods.

Example 3.7

Now consider again the rolling of a die with two new events:
C: A multiple of 2 is obtained.
D: A multiple of 3 is obtained.

So the probability of the event D is $\frac{2}{6}$, while the conditional probability of D, given that C has occurred, is $\frac{1}{3}$. Since $\frac{2}{6} = \frac{1}{3}$, we can deduce that the two events are independent: knowing that an even number has occurred does not make it any more, or less, likely that some multiple of 3 has occurred.

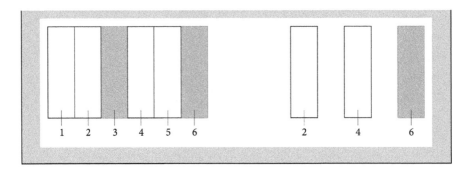

> The previous example illustrated that events can be mutually independent _without_ being mutually exclusive. However, mutually exclusive events (with positive probabilities) can _never_ be mutually independent.

If events E and F can both occur, but are mutually exclusive, then $P(E|F) = 0$, Since the probability of E occurring is affected by knowledge that F has occurred, it follows that E and F are not independent events.

3.4 The total probability theorem

The total probability theorem amounts to the statement that the whole is the sum of its parts. A simple illustration of the general idea is provided by Figure 3.6.

It may help to think of B as a potato, with A being a five-compartment potato slicer. The five compartments have different widths because the events $A_1, A_2, \cdots, A_5$ are not equally likely.

The potato may be sliced into as many as five parts, though in this case there are just four: the part falling in the first compartment, the part falling

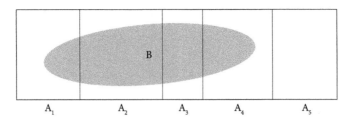

Figure 3.6 An event is the sum of its parts.

in the second compartment, and the parts falling into the third and fourth compartments. The algebraic equivalent of reassembling the potato from its component parts is:

$$P(B) = P(B \cap A_1) + P(B \cap A_2) + \cdots + P(B \cap A_5).$$

In this case, since the fifth compartment is empty, $P(B \cap A_5) = 0$.
Using Equation (3.4) this becomes:

$$P(B) = \{P(B|A_1) \times P(A_1)\} + \{P(B|A_2) \times P(A_2)\} + \cdots + \{P(B|A_5) \times P(A_5)\}.$$

Generalizing to the case where A has n categories we get the theorem:

$$P(B) = \sum_{i=1}^{n} P(B \cap A_i) = \sum_{i=1}^{n} P(B|A_i) \times P(A_i). \qquad (3.5)$$

Example 3.8

It is known that 40% of students are good at Physics. Of those students, 80% are also good at Mathematics. Of those students who are not good at Physics, only 30% are good at Mathematics. We wish to find the overall proportion of students that are good at Mathematics.
   The information is collected together in Figure 3.7:

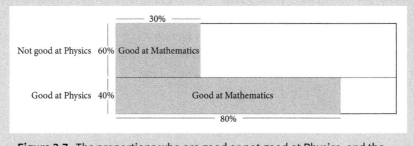

Figure 3.7 The proportions who are good or not good at Physics, and the proportions of those groups that are good at Mathematics.

Here we take A_1 to be the event 'Good at Physics', with A_2 being 'Not good at Physics'. We require $P(B)$, where B is the event 'Good at Mathematics'. We are told that $P(A_1) = 0.4$, which implies that $P(A_2) = 0.6$, since these are the only possibilities. We are also told that $P(B|A_1) = 0.8$ and $P(B|A_2) = 0.3$, Using Equation (3.5), we have

$$
\begin{aligned}
P(B) &= \{P(A_1) \times P(B|A_1)\} + \{P(A_2) \times P(B|A_2)\} \\
&= (0.4 \times 0.8) + (0.6 \times 0.3) \\
&= 0.32 + 0.18 = 0.50
\end{aligned}
$$

Thus half the students do well in Mathematics.

3.5 Bayes' theorem

In introducing the idea of conditional probability we effectively asked the question:

'Given that event A has occurred in the past, what is the probability that event B will occur?'

We now consider the following 'reverse' question:

'Given that the event B has just occurred, what is the probability that it was preceded by the event A?'.

In the previous section we imagined a potato being sliced into segments. Now suppose those segments are placed into a black bag and a skewer pierces one segment. The probability of a particular segment being skewered will be equal to the size of that segment when regarded as a proportion of the whole potato. The biggest piece will have the highest probability.

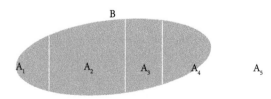

Figure 3.8 The biggest segment is the most likely to occur.

Reverting to the general case, the whole potato represented P(B), with segment i representing $P(B \cap A_i)$. So, if we know that the event B has occurred, then the probability that that was a consequence of event A_i having occurred is given by $P(B \cap A_i)/P(B)$. Using Equation (3.5), we have **Bayes' theorem:**[3]

$$P(A_i|B) = \frac{P(B \cap A_i)}{P(B)} = \frac{P(B|A_i) \times P(A_i)}{\sum_{i=1}^{n} P(B|A_i) \times P(A_i)}. \qquad (3.6)$$

Example 3.9

A test has been devised to establish whether a patient has a particular disease. If the patient has the disease, then there is a 2% chance that the test will not detect it (a **false negative**). If the patient does not have the disease, there is nevertheless a 2% chance that the test will report that the patient has the disease (a **false positive**). Suppose that 4% of the population have the disease. The question of interest to the patient is:

If I am diagnosed as having the disease, what is the probability that I do indeed have it?

It is easy to suppose that, since the test is 98% accurate, the probability that the patient has the disease must be 98%. This is not the case. We start by arranging the information in a table. For convenience, suppose that we have a population of 10,000 people. Then 4% = 400 have the disease, and 9600 do not. Of the 400 with the disease, 2% = 8, test negative and 392 test positive. Of the 9600 without the disease, 2% = 192, test positive. These numbers are set out in the table:

	Positive	Negative	Total
Has disease	392	8	400
Does not have disease	192		9600
Total	584		10000

---

[3] The Reverend Thomas Bayes (1701–1761) was a Nonconformist minister in Tunbridge Wells, Kent. He was elected a Fellow of the Royal Society in 1742. The theorem was contained in an essay that did not appear until after his death and was largely ignored at the time.

We see that of the 584 diagnosed positive, only 392 actually have the disease. Formally, define the events B, A_1, and A_2 as follows:

B: Tests positive

A_1: Has the disease.

A_2: Does not have the disease.

We need to calculate $P(A_1|B)$. To evaluate this we first calculate $P(B)$. Using Equation (3.5):

$$P(B) = 0.04 \times 0.98 + 0.96 \times 0.02 = 0.0392 + 0.0192 = 0.0584.$$

Using Equation (3.6), we find that

$$P(A_1|B) = \frac{0.0392}{0.0584} = 0.67.$$

Despite the high accuracy of the test, only about two-thirds of patients who test positive actually have the disease.

Reminder: Probabilities never exceed 1 and are never negative.

4

Probability distributions

4.1 Notation

Conventionally, random variables are denoted using capital letters: X, Y, ...,
whereas observations and the values that random variables might take are
denoted using lower-case letters: x, y, ... So we might write $P(X = x)$ to
mean 'the probability that the random variable X takes the value x'. In this
chapter we concentrate on situations where an appropriate summary of the
data might be a tally chart, and an appropriate representation would be a bar
chart (rather than a histogram).

Suppose that X is a discrete random variable that can take the values x_1, x_2,
..., x_m and no others. Since X must take some value:

$$P(X = x_1) + P(X = x_2) + \cdots + P(X = x_m) = 1.$$

The sizes of $P(X = x_1)$, $P(X = x_2)$, ..., show how the total probability of 1
is *distributed* amongst the possible values of X. The collection of these values
therefore defines a **probability distribution**.

4.2 Mean and variance of a probability distribution

Equation (2.2) gave the sample mean as,

$$\bar{x} = \frac{1}{n} \sum_{j=1}^{m} f_j x_j,$$

where n is the sample size, $x_1, x_2, ..., x_m$ are the distinct values that occur in
the sample, and f_j is the frequency with which the value x_j occurs.

Taking the $\frac{1}{n}$ inside the summation gives:

$$\bar{x} = \sum_{j=1}^{m} \frac{f_j}{n} x_j.$$

Data Analysis: A Gentle Introduction for Future Data Scientists. Graham Upton and Dan Brawn, Oxford University Press.
© Graham Upton and Dan Brawn (2023). DOI: 10.1093/oso/9780192885777.003.0004

The ratio $\frac{f_j}{n}$ is the proportion of the sample for which X equals x_j. Denoting this proportion by p_j, the formula for the sample mean becomes:

$$\bar{x} = \sum_{j=1}^{m} p_j x_j,$$

In the same way, the sample variance, given by Equation (2.3), can be written as:

$$\frac{n}{n-1} \sum_{j=1}^{m} p_j (x_j - \bar{x})^2.$$

Now imagine the sample increasing in size to include the entire population. Two consequences are:

1. Every possible value of X will occur at least once, and
2. The proportion of the 'sample' taking the value x_j will then be equal to the probability that a randomly chosen member of the population has the value x_j.

The population mean, which is usually denoted by μ, is therefore given by:

$$\mu = \sum P(X = x)x, \tag{4.1}$$

where the summation is over all possible values of x. As n increases, the ratio $n/(n-1)$ comes ever closer to 1, while $\bar{x}$ comes ever closer to μ. The population variance, usually denoted by σ^2, is therefore given by:

$$\sigma^2 = \sum P(X = x)(x - \mu)^2. \tag{4.2}$$

A little bit of algebra gives the equivalent formula:

$$\sigma^2 = \sum P(X = x)x^2 - \mu^2. \tag{4.3}$$

We can use whichever is the more convenient; we will get the same answer either way!

Example 4.1

As an example, consider the result of rolling a fair die, numbered in the usual way from 1 to 6. With six equally likely possibilities, the probability of each is $\frac{1}{6}$. So

$$\mu = \frac{1}{6} \times 1 + \frac{1}{6} \times 2 + \cdots + \frac{1}{6} \times 6 = \frac{1}{6} \times 21 = \frac{7}{2} = 3.5.$$

Using Equation (4.2), the variance is:

$$\sigma^2 = \frac{1}{6} \times (1 - 3.5)^2 + \frac{1}{6} \times (2 - 3.5)^2 + \cdots + \frac{1}{6} \times (6 - 3.5)^2 = \frac{35}{12}.$$

This is a rather messy calculation. Using Equation (4.3) is more straightforward, since:

$$\sum P(X = x)x^2 = \frac{1}{6} \times (1 + 4 + 9 + 16 + 25 + 36) = \frac{91}{6},$$

so that:

$$\sigma^2 = \frac{91}{6} - \left(\frac{7}{2}\right)^2 = \frac{35}{12}.$$

4.3 The relation between sample and population

In the previous section we used the formulae for sample mean and variance to deduce the equivalent formulae for populations. We now look at two contrasting examples.

Example 4.2

As an example of the relation between the sample and population statistics, we again use the simulated die-rolling results given in Table 3.1 and included now in Table 4.1:

Table 4.1 Extended results of 'rolling a six-sided die'.

n	1	2	3	4	5	6	$\bar{x}$	s^2
	\multicolumn{6}{c}{Proportion showing a particular face}		
6	0.500	0.167	0.000	0.167	0.167	0.000	2.333	3.067
60	0.133	0.283	0.100	0.183	0.167	0.133	3.367	2.779
600	0.162	0.195	0.160	0.153	0.163	0.167	3.462	2.940
6000	0.168	0.174	0.166	0.172	0.165	0.156	3.460	2.867
60,000	0.169	0.166	0.167	0.165	0.168	0.165	3.491	2.923
600,000	0.167	0.167	0.167	0.167	0.166	0.167	3.499	2.920

Extending the simulation to 600,000 'rolls of the die', we can see that all the sample proportions are very close to their common population probability, 1/6. In the same way we see that the sample mean is close to the population mean, 3.5. The sample variance is similarly close to the population variance, σ^2 (which, in the previous example, we found to be 35/12 = 2.917).

Example 4.3

Huntington's disease (HD) is an illness caused by a faulty gene that affects the nervous system. A study of UK adults aged over 21, found that, in 2010, there were about 120 HD cases per million population. With 40 million adults in the UK this implies that there were about 4800 individuals with HD.

Suppose that we sample in a fashion that makes every one of the 40 million adults equally likely to be, selected. How big would the sample need to be, if the aim is to include five individuals suffering with HD?

We can get an idea by using the computer to simulate selecting n individuals from a population of 40 million. If we repeat this 100 times for each value of n, then we can estimate the probability of getting at least five HD individuals in a sample of that size. The results are shown in Table 4.2

Table 4.2 Numbers of HD suffers in 100 samples of size n from a population of 40 million individuals.

n	Number with HD (in sample of n)					
---	0	1	2	3	4	≥ 5
1000	83	17	0	0	0	0
2000	82	18	3	0	0	0
5000	44	37	14	7	4	1
10000	31	34	23	10	6	2
20000	10	19	29	19	13	10
50000	0	3	4	7	14	72

The table shows that a sample of a few thousand individuals is certainly insufficient if the aim is to find 5 HD individuals. Indeed, even with a sample of 20,000 individuals there is a 10% chance of not finding a single HD patient. For reasonable confidence one would need to sample some 50,000 individuals. In practice, of course, the sampling would be targeted, using medical records.

The table demonstrates the extent to which a sample can misrepresent the population: for example, with $n = 50,000$, on four of the 100 trials a total of 11 HD patients were encountered. This corresponds to a rate of 220 per million, which is nearly double the true incidence rate (120). At the other extreme, on three trials only a single HD sufferer was found: a rate of just 20 per million.

- *The so-called **laws of large numbers** provide mathematical assurances that as n increases, so the difference between the sample mean ($\bar{x}$) and the population mean (μ) is sure to ultimately reduce to zero.*
- *Unless the distribution is very skewed the majority of values will usually lie in the range ($\mu - 2\sigma, \mu + 2\sigma$), where μ and σ are the population mean and standard deviation.*

With a very skewed distribution a sample may need to be very large indeed if we require reliable information.

4.4 Combining means and variances

If X and Y are independent random variables, then writing Var for variance, there are some simple rules:

1. The mean of $X + Y$ equals the mean of X plus the mean of Y.
2. $\text{Var}(X + Y)$ equals $\text{Var}(X)$ plus $\text{Var}(Y)$.
3. The mean of aX equals a times the mean of X.
4. $\text{Var}(aX)$ equals a^2 times $\text{Var}(X)$.

Using rules 2 and 4 we find that:

$$\text{Var}(X - Y) = \text{Var}(X) + \text{Var}(Y), \tag{4.4}$$

which shows that the difference between X and Y has the same variance as their sum $X + Y$.

Beware: $Var(X - Y) \neq Var(X) - Var(Y)$.

Rules 1 to 4 extend naturally when combining information on more than two variables.

A case of particular interest occurs when we take a sample of n independent observations from a population. We can think of each member of the sample as being an observation on its own random variable. We could call these variables 'The first sample value', X_1, 'The second sample value', X_2, and so on up to X_n. Since they all come from the same population, each random variable has the same mean, μ, and the same variance, σ^2. Thus:

5. $(X_1 + X_2 + \cdots + X_n)$ has mean $n\mu$.
6. $(X_1 + X_2 + \cdots + X_n)$ has variance $n\sigma^2$.

Using these results together with rules 3 and 4, and noting that:

$$\bar{X} = \frac{1}{n}(X_1 + X_2 + \cdots + X_n),$$

we have the results:

7. $\bar{X}$ has mean $\frac{1}{n} \times n\mu = \mu$ (from rules 3 and 5).
   This shows that the sample mean, $\bar{x}$, is an unbiased estimate of the population mean, μ.
8. $\bar{X}$ has variance $\left(\frac{1}{n}\right)^2 \times n\sigma^2 = \frac{\sigma^2}{n}$ (from rules 4 and 6).
   This shows that increasing the sample size, n, reduces the variance of the estimate of the population mean and hence makes it more accurate. Specifically, variance reduces inversely with sample size.

Taken together, results 7 and 8 underpin the laws of large numbers (see Section 4.3). Result 8 tells us that as n increases, so $\bar{X}$ becomes less variable. Result 7 shows that the value about which it is decreasingly variable is μ. In other words, as n increases, so (subject to random variation) the value of the sample mean converges to the population mean.

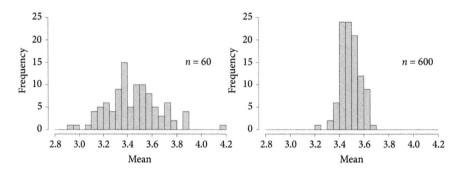

Figure 4.1 As the sample size increases so the variability of the sample mean decreases and it becomes an increasingly precise estimate of the population mean.

In the context of determining the sample mean from rolls of a fair die numbered from 1 to 6, Figure 4.1 shows the effect of increasing the sample size from 60 to 600. Each diagram summarizes the means of 100 samples.

4.5 Discrete uniform distribution

Here the random variable X is equally likely to take any of k values $x_1, x_2, \ldots,$ x_k, and can take no other value. The distribution is summarized by:

$$P(X = x_i) = \begin{cases} \frac{1}{k} & i = 1, 2, \ldots, k, \\ 0 & \text{otherwise.} \end{cases} \tag{4.5}$$

Example 4.4

Suppose that the random variable X denotes the face obtained when rolling a fair six-sided die with faces numbered 1 to 6. The probability distribution is given by:

$$P(X = x) = \begin{cases} \frac{1}{6} & x = 1, 2, \ldots, 6, \\ 0 & \text{otherwise.} \end{cases}$$

4.6 Probability density function

With a continuous variable, probability is related to a range of values rather than any single value. For continuous data, we used histograms (Section 2.4) to illustrate the results. With a small amount of data, the histogram is inevitably rather chunky, but as the amount of data increases, so the histogram can develop a more informative outline (see Figure 4.2).

The peak of a histogram corresponds to the range of values where the values are most dense. Ultimately, as we move from large sample to population, so we move from proportion to probability. At that point we need only the outline, which is a function of the values recorded. This is called the **probability density function**, which is commonly shortened to **pdf**, and is designated by f(). For the situation leading to the histograms in Figure 4.2, the probability density function is that illustrated in Figure 4.3.

The minimum value for a pdf is zero. Values for which f is zero are values that will never occur.

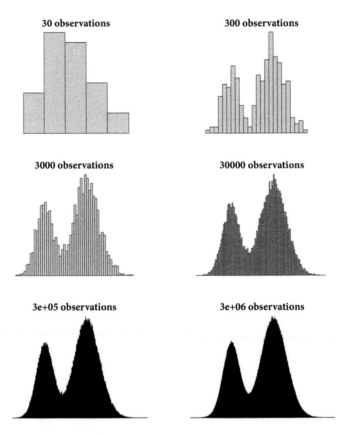

Figure 4.2 As the sample size increases so the histogram becomes more informative concerning the values in the population being sampled.

Closely related[1] to the probability density function is the **cumulative distribution function** (cdf), F(). For a value x, the value taken by the cdf, is the probability that the random variable X takes a value less than or equal to x:

$$F(x) = P(X \leq x). \tag{4.6}$$

For every distribution (discrete or continuous), $F(-\infty) = 0$ and $F(\infty) = 1$. Figure 4.2 showed how, as the amount of data increases, so the outline of the histogram increasingly resembles the probability density function describing the values in the population. In the same way, successive plots (for increasing amounts of data) lead the outline of a cumulative frequency diagram (Section 2.5) to resemble the cumulative distribution function. In the case of a discrete random variable the cdf will be a step diagram (Section 2.6).

---

[1] For a continuous random variable, the cumulative distribution function, F(x), is the integral of the density function, f(x).

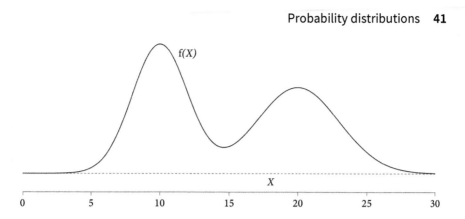

Figure 4.3 The probability density function that was randomly sampled to obtain the data for the histograms of Figure 4.2.

4.7 The continuous uniform distribution

This is the simplest distribution to describe, as it states that all the values in some range (from a to b, say), are equally likely to occur, with no other values being possible.[2]

> *The uniform distribution, also called the* **rectangular distribution**, *has mean* $= \frac{1}{2}(a + b)$ *and variance* $= \frac{1}{12}(b - a)^2$.

> *When a measurement is rounded to a whole number, the associated* **rounding error** *has a uniform distribution in the range –0.5 to 0.5.*

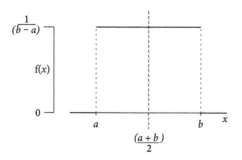

Figure 4.4 A uniform distribution between the values a and b.

---

[2] The pdf for a distribution that is uniform between the values a and b, with no other values being possible, is given by f(x) $= \frac{1}{b-a}$ for $a < x < b$ (see Figure 4.4).

Example 4.5

The distance between points A and B is recorded as 4, while that between B and C is recorded as 6. Distances have been rounded to whole numbers. The best estimate of the distance between A and C via B is therefore 10, but the question is 'How accurate is that estimate?'.

The true distance between A and B is somewhere between 3.5 and 4.5, with no value being more likely than any other. This is therefore an observation from a uniform distribution mean 4 and variance $(4.5 - 3.5)^2/12 = 1/12$. Similarly, the distance between B and C has mean 6 and variance 1/12.

Using the results of Section 4.4, the means and the variances simply add together so the entire journey has mean $4 + 6 = 10$ and variance equal to $1/12 + 1/12$ which corresponds to a standard deviation for the distance of the entire journey as $1/\sqrt{6} \approx$ 0.408 units.

5
Estimation and confidence

5.1 Point estimates

A **point estimate** is a numerical value, calculated from a set of data, which is
used as an estimate of an unknown parameter in a population. The random
variable corresponding to an estimate is known as the **estimator**. The most
familiar examples of point estimates are:

sample mean, $\bar{x}$	used to estimate the population mean, μ
sample proportion, r/n,	used to estimate the population proportion, p;
sample variance, s^2	used to estimate the population variance, σ^2.

These three estimates are very natural ones and they also have a desirable
property: they are **unbiased**. That is to say that their long-term average values
are precisely equal to the population values.

5.1.1 Maximum likelihood estimation (mle)

This is the method most often used by statisticians for the estimation of
quantities of interest.

The **likelihood** reflects the probability that a future set of n observations
have precisely the values observed in the current set. In most cases, while
the form of the distribution of X may be specified (e.g., binomial, normal,
etc.), the value of that distribution's parameter or parameters (e.g., p, μ, σ^2,
etc.) will not be known. The likelihood will therefore depend on the values of
these unknown parameter(s). A logical choice for the unknown parameter(s)
would be those value(s) that maximize the probability of recurrence of the
observations: this is the principle of **maximum likelihood**.

5.2 Confidence intervals

A point estimate is just that: an estimate. We cannot expect it to be exactly
accurate. However, we would be more confident with an estimate based on a

Data Analysis: A Gentle Introduction for Future Data Scientists. Graham Upton and Dan Brawn, Oxford University Press.
© Graham Upton and Dan Brawn (2023). DOI: 10.1093/oso/9780192885777.003.0005

large sample than one based on a small sample (assuming that both samples were unbiased).

A confidence interval, which takes the general form (Lower bound, Upper bound), quantifies the uncertainty in the value of the point estimate.

5.3 Confidence interval for the population mean

Of all the properties of a population, the average value is the one that is most likely to be of interest. Underlying the calculation of the confidence interval for the population mean is the so-called normal distribution (also known as the Gaussian distribution) which we now introduce.

5.3.1 The normal distribution

The normal distribution describes the very common situation in which very large values are rather rare, very small values are rather rare, but middling values are rather common. Figure 5.1 is a typical example of data displaying a normal distribution. The data refer to verbal IQ scores of schoolchildren in the Netherlands.[1]

The distribution is unimodal and symmetric with two parameters, μ (the mean) and σ^2 (the variance).[2] As a shorthand we refer to a $N(\mu, \sigma^2)$

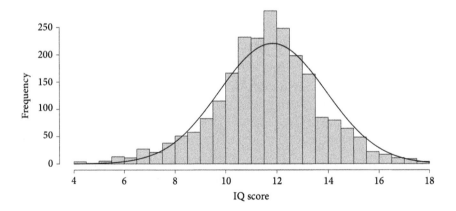

Figure 5.1 The verbal IQ scores of 2287 eighth-grade pupils (aged about 11) in schools in the Netherlands, together with a normal distribution having the same mean and variance.

[1] The data are used in the form presented by the authors of the book in which the data first appeared.

[2] The probability density function is given by $f(x) = \frac{1}{\sigma\sqrt{2\pi}} \exp\left(-\frac{(x-\mu)^2}{2\sigma^2}\right)$.

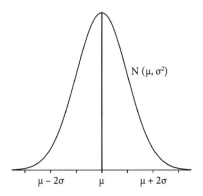

$N(\mu, \sigma^2)$

$\mu - 2\sigma$ μ $\mu + 2\sigma$

Figure 5.2 The probability density function for a normal distribution with mean μ and variance σ^2.

distribution. Because of the symmetry, the mean is equal to both the mode and the median. The distribution is illustrated in Figure 5.2 where the x-axis is marked in intervals of σ.

Roughly 95% of values lie within about two standard deviations from the mean.

The normal distribution with mean 0 and variance 1 is described as the **standard normal distribution** and you will often see it denoted as $N(0,1)$.

5.3.2 The Central Limit Theorem

This is the reason why the normal distribution is so frequently encountered.

If each of the independent random variables $X_1, X_2, \ldots, X_n$ has the same distribution (with finite mean and variance), then, as n increases, the distribution of both their sum and their average increasingly resembles a normal distribution.

The importance of the **Central Limit Theorem** is because:

- The common distribution of the X-variables is not stated—it can be almost *any* distribution.
- In most cases, the resemblance to a normal distribution holds for remarkably small values of n.

As an example, Figure 5.3 shows results for observations for a uniform distribution. Each histogram reports 1000 results for the original distribution, for the means of pairs of observations, for the means of four observations, and for the means of eight observations. As the group size increases so the means become increasingly clustered in a symmetrical fashion about the mean of the population.

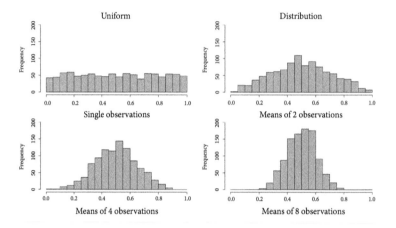

Figure 5.3 The distributions of means of observations from a uniform distribution.

As a second example, Figure 5.4 shows histograms for successive averages of observations from the opposite of a normal distribution: a situation where central values are uncommon, and most values are very large or very small.

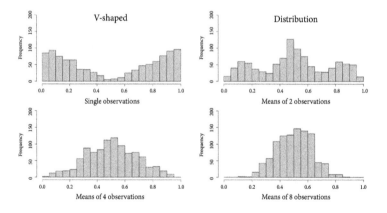

Figure 5.4 The distributions of means of observations from a V-shaped distribution.

Once we start looking at averages of even as few as $n = 2$ observations, a peak starts to appear.

As a final example (Figure 5.5) we look at a very skewed distribution. The skewness has almost vanished by the time we are working with means of eight observations.

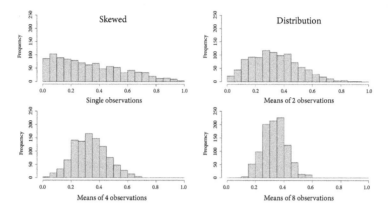

Figure 5.5   The distributions of means of observations from a skewed distribution.

Suppose a distribution has mean μ and standard deviation σ. The mean of a random sample of n observations from this distribution can be regarded as being an observation from a normal distribution with mean μ and standard deviation $\sigma/\sqrt{n}$.

It is clear that the practical consequences of the Central Limit Theorem were understood well before the time of Laplace. A sixteenth-century German treatise on surveying instructs the surveyor to establish the length of a rood (a standard unit of length) in the following manner:

'Stand at the door of a church on a Sunday and bid 16 men to stop, tall ones and small ones, as they happen to pass out when the service is finished; then make them put their left feet one behind the other, and the length thus obtained shall be a right and lawful rood to measure and survey the land with, and the 16th part of it shall be a right and lawful foot.'

5.3.3 Construction of the confidence interval

The normal distribution was introduced because the confidence interval for a mean relies upon it. We noted earlier that roughly 95% of values lie within about two standard deviations of the centre of a normal distribution, which is at μ. So we know that the observed sample mean, $\bar{x}$, will generally be within two standard deviations of μ. Equally, μ will be within two standard deviations of $\bar{x}$.

Usually the value of σ^2, is unknown, but, in a large sample, a good approximation[3] will be provided by the sample variance, s^2. So an approximate 95% confidence interval for a population mean is:

$$\left(\bar{x} - 2\frac{s}{\sqrt{n}}, \quad \bar{x} + 2\frac{s}{\sqrt{n}}\right). \tag{5.1}$$

Example 5.1

A machine fills 500 ml containers with orange juice. A random sample of 10 containers are examined in a laboratory, and the amounts of orange juice in each container were determined correct to the nearest 0.05 ml. The results were as follows:

503.45, 507.80, 501.10, 506.45, 505.85, 504.40, 503.45, 505.50, 502.95, 507.75.

Since the total volume of juice in a container can be thought of as the sum of the volumes of large numbers of juice droplets, it is reasonable to assume that these are observations from a normal distribution.

The sample mean and standard deviation are 504.87 and 2.18, respectively, so that an approximate 95% confidence interval for the population mean is $504.87 \pm 2 \times \frac{2.18}{\sqrt{10}}$, which gives (503.5, 506.3).

For a single bottle ($n = 1$) the approximate interval is much wider at (500.5, 509.2), which excludes 500 ml.

5.4 Confidence interval for a proportion

Once again the computer can be relied upon to do the calculations. However, this time the user may obtain slightly different answers depending on the choice of computer routine. To discover why this is, we first introduce the underlying distribution.

---

[3] Statisticians would observe that a more accurate interval would replace the multiplier 2 by an appropriate value from a t-distribution with $n - 1$ degrees of freedom.

5.4.1 The binomial distribution

The word 'binomial' means having two terms. For this distribution to be appropriate the following must be true:

- There is a fixed number, n, of independent identical trials.
- Each trial results in either a 'success' or a 'failure' (the two terms).
- The probability of success, p, is the same for each trial.

The probability of obtaining exactly r successes in the n trials is given by:

$$P(X = r) = \binom{n}{r} p^r (1 - p)^{n-r}, \tag{5.2}$$

where

$$\binom{n}{r} = \binom{n}{n-r} = \frac{n!}{r!(n-r)!} \tag{5.3}$$

is the number of ways of choosing r out of n, and:

$$r! = r \times (r-1) \times \cdots \times 1, \text{ for } r > 0 \qquad \text{with} \qquad 0! = 1. \tag{5.4}$$

Two examples with $n = 10$ are illustrated in Figure 5.6.

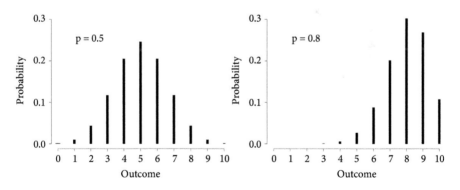

Figure 5.6 Two binomial distributions with $n = 10$. The distribution is symmetric when $p = 0.5$.

The binomial distribution has mean np and variance $np(1 - p)$.

5.4.2 Confidence interval for a proportion (large sample case)

When n is large the shape of the binomial distribution resembles that of a normal distribution to the extent that the latter can be used as an approximation that avoids heavy computations.

Suppose that we have r successes in n trials, so that the point estimate, $\hat{p}$, is given by $\hat{p} = r/n$. Using the normal approximation, r is an observation from a normal distribution with $\mu = np$ and $\sigma^2 = np(1-p)$. Thus $\hat{p}$ has mean $np/n = p$ and variance (using Rule 4 of Section 4.4) $p(1-p)/n$. Substituting $\hat{p}$ for p, leads to an approximate 95% confidence interval for p as:[4]

$$\left(\hat{p} - 2\sqrt{\frac{\hat{p}(1-\hat{p})}{n}}, \quad \hat{p} + 2\sqrt{\frac{\hat{p}(1-\hat{p})}{n}} \right). \tag{5.5}$$

Example 5.2

As an example suppose that $r = 50$ and $n = 250$, so that $\hat{p} = 0.2$. The approximate 95% confidence interval is $(0.15, 0.25)$.

5.4.3 Confidence interval for a proportion (small sample)

With small samples the normal approximation to the binomial is no longer appropriate. We need to work directly with the binomial distribution itself. However, this is not straightforward and there is considerable discussion in the statistical literature on how best to proceed. Different approaches lead to different bounds. Fortunately, the data scientist will rarely encounter small samples!

5.5 Confidence bounds for other summary statistics

5.5.1 The bootstrap

This approach depends on the ability of the modern computer to carry out a huge number of calculations very quickly. No assumptions are made about the underlying population distribution, and no use is made of the Central Limit Theorem. Instead, in response to the question 'What *is* known about

---

[4] Depending upon the package used, the method may be attributed to either Laplace or Wald.

the distribution?', the answer is 'The values $x_1, x_2, ..., x_n$ can occur'. Assume for clarity that all n values are different, then those n values are apparently equally likely to appear. The assumed distribution is therefore given by:

$$P(X = x_i) = \frac{1}{n}, \qquad (i = 1, 2, ..., n). \qquad (5.6)$$

Any intermediate values are supposed to have zero probability of occurrence.

The next question is 'If this is the distribution, and we take n observations from the distribution, *with replacement* after each observation, then what sample values might have been obtained?' This is called **resampling**. Note that each new sample contains only the values from the original sample, but some of those values may appear more than once, while some will not appear at all. As an example, suppose that the data vector consists of the numbers 0 to 9. Successive samples of 10 observations might be:

```
Sample 1:  5  5  3  9  7  0  7  7  3  8
Sample 2:  3  5  8  1  3  8  4  1  2  6
Sample 3:  1  7  7  7  5  1  1  9  8  0
```

A key feature is that not only are the samples different from one another but they have potentially different means, medians, variances, etc.

Suppose that for some reason we are interested in the median. We not only have an observed value but also the possibility, by resampling, to find the other values that might have occurred. We would use at least 100 resamples.

Remember that the original sample will not exactly reproduce the population from which it is drawn. Since the resampling is based on the sample rather than the unknown population, the inevitable biases in the sample will be manifest in the bootstrap distribution.

If the true distribution is known, or can be approximated, then methods based on that distribution should be used in preference to the bootstrap.

Example 5.3

To illustrate the procedure we need some data. Here are 30 observations (randomly generated from a uniform distribution with a range from 0 to 500):

457	469	143	415	321	260	368	67	328	353
229	360	467	128	231	470	489	59	237	280
452	69	494	473	41	257	195	453	223	418

Suppose that we are interested in the **coefficient of variation**, cv, which is a quantity that compares the variability of a set of data with its typical value by dividing the standard deviation by the mean. In this case the observed value is reported (by R) as 0.4762557.

We will now use the bootstrap with 1,000,000 resamples to find a 95% confidence interval for this value. The computer gives the apparently highly accurate result `(0.3403718, 0.6097751)`.

Do not be persuaded by the large number of resamples and the seven significant figures into thinking that high precision has thereby been achieved. Even with such a huge number of resamples, repeat runs will produce different values. In this case, two further runs gave the bounds `(0.3405424, 0.6099814)` and `(0.3404129, 0.6099990)` which are very similar, but not identical. A reasonable report would be that the 95% bounds are (0.34, 0.61).

Since we generated the data from a known distribution this means we also know the true value of cv. The distribution (see Section 4.7) has mean 250 and standard deviation $500/\sqrt{12}$ so that the true cv is 0.58. This is much higher than the observed value (0.4762557); it only just lies within the confidence interval. This serves as a reminder that relatively small samples may not accurately reflect the population from which they came.

5.6 Some other probability distributions

5.6.1 The Poisson and exponential distributions

5.6.1.1 Poisson process
When every point in space (or time, or space-time) is equally likely to contain an event, then the events are said to occur at random. In mathematical terms this is described as a Poisson process.[5]

Figure 5.7 illustrates a temporal Poisson process, with the passing of time indicated by the position on the line from left to right. The time line is subdivided into 10 equal lengths. There is a total of 55 events, but these are not at all evenly distributed, with, by chance, a concentration of events in the first tenth.

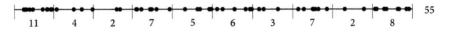

| 11 | 4 | 2 | 7 | 5 | 6 | 3 | 7 | 2 | 8 | 55 |

Figure 5.7 Events occurring at random points on the time line: an example of a temporal Poisson process.

[5] Siméon Denis Poisson (1781–1840) was a French mathematician whose principal interest lay in aspects of mathematical physics. His major work on probability was entitled *Researches on the Probability of Criminal and Civil Verdicts*. In this long book (over 400 pages) only about one page is devoted to the derivation of the distribution that bears his name.

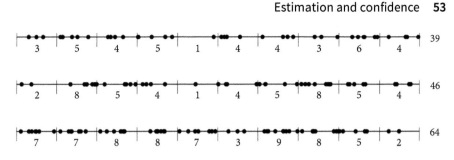

Figure 5.8 Three further examples of a temporal Poisson process.

This mixture of gaps and clumps is typical of events happening at random. When the events are catastrophic (plane crashes, wildfires, etc.) the clumping of random events should be borne in mind. Whilst there may be some common cause for the catastrophes, this need not be the case.

Figure 5.8 gives three further examples of a Poisson process. All four examples refer to the same length of time, with the random events occurring at the same rate throughout. The differences between the total numbers of events is a further example of the way in which randomness leads to clusters of events.

5.6.1.2 Poisson distribution

For events following a Poisson process, the Poisson distribution[6] describes the variation in the numbers of events in equal-sized areas of space or intervals of time. Two examples of Poisson distributions are illustrated in Figure 5.9.

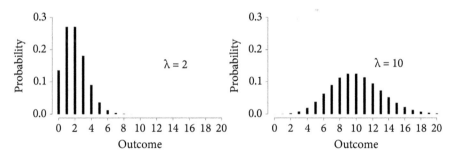

Figure 5.9 The Poisson distribution becomes more symmetric as the mean, λ, increases.

> The Poisson distribution has mean and variance equal to one another.

---

[6] For events occurring at a rate of λ per unit area (or time), the probability that a unit area contains exactly x events is given by $P(X = x) = \frac{\lambda^x e^{-\lambda}}{x!}$, $(x = 0, 1, 2, ...)$, where e is the exponential function, and the quantity $x!$ was given by Equation (5.4).

Example 5.4

Figure 5.10 illustrates a spatial Poisson process. The event positions in two dimensions were generated using pairs of random numbers. A summary of the counts in 100 sub-regions is given in Table 5.1.

1	2	1	0	1	2	1	1	1	1
0	5	0	1	0	2	2	0	1	2
2	0	2	0	1	1	2	1	2	2
1	4	1	1	1	1	0	1	0	1
2	2	1	0	0	1	1	2	1	0
1	1	0	1	0	1	1	0	0	3
0	2	1	1	4	0	0	2	3	1
2	4	3	2	0	2	2	2	1	2
2	1	2	1	0	3	3	1	1	2
2	0	4	0	2	0	0	0	1	0

Figure 5.10 The left-hand diagram illustrates the positions of 125 randomly positioned points. The right-hand diagram reports the counts in the 100 sub-regions.

Table 5.1 A summary table of the numbers of events in the 100 sub-regions of Figure 5.10.

Number of events, x, in a sub-region	0	1	2	3	4	5
Number of sub-regions containing x events	27	37	26	5	4	1

If the events are distributed at random in space then the numbers of events in each sub-region will be observations from a Poisson distribution. Since Poisson distributions have mean equal to variance, a useful check for randomness is provided by comparing the sample mean with the sample variance. In this case the sample mean is 1.25 and the sample variance is 1.20, so the pattern is consistent with randomness.

5.6.1.3 The exponential distribution

This distribution refers to the lengths of the intervals between events occurring at random points in time.[7] Two examples of the distribution are illustrated in Figure 5.11.

> The exponential distribution has mean $1/\lambda$ and variance $1/\lambda^2$, with $P(X > x) = \exp(-\lambda x)$.

---

[7] It has probability density function given by $f(x) = \lambda e^{-\lambda x}$ $x > 0$.

Figure 5.12 shows a histogram of the intervals between the random events shown in Figure 5.7. This is an example of data from an exponential distribution.

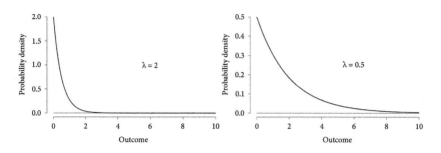

Figure 5.11 Two exponential distributions.

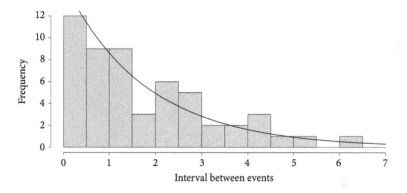

Figure 5.12 Histogram of the intervals between the random events of Figure 5.7, together with the fitted exponential distribution.

5.6.2 The Weibull distribution

The Weibull[8] distribution is an example of an **extreme value distribution**. It is often used to model the lifetimes of systems of components. Its probability density function involves two parameters, k and λ. As Figure 5.13 demonstrates, the shape of the distribution is critically dependent on the value of k, with the case $k = 1$ corresponding to the exponential distribution.

[8] Ernst Hjalmar Wallodi Weibull (1887–1979) was a Swedish engineer. His career was in the Royal Swedish Coast Guard, where he studied component reliability. In 1972 the American Society of Mechanical Engineers awarded him their Gold Medal and cited him as 'a pioneer in the study of fracture, fatigue and reliability'.

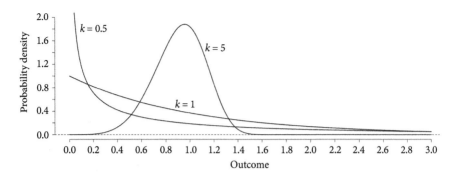

Figure 5.13 Three Weibull distributions, each with $\lambda = 1$.

The exponential distribution assumed that events such as failures occurred at a steady rate over time. However, often there is a small proportion of components that fail almost immediately, while for other components the failure rate increases as the components get older. The result is a characteristic 'bathtub' curve. The example shown in Figure 5.14 was created by mixing the three Weibull distributions given in Figure 5.13.

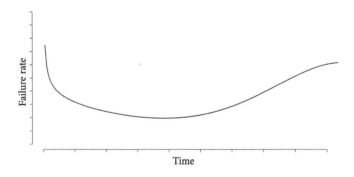

Figure 5.14   A bathtub curve showing how the failure rate typically varies over time.

5.6.3 The chi-squared (χ^2) distribution

'Chi' is the Greek letter χ, pronounced 'kye'. The distribution is continuous[9] and has a positive integer parameter d, known as the degrees of freedom, which determines its shape. Some examples of chi-squared distributions are illustrated in Figure 5.15.

---

[9] The probability density function $f(x) \propto \exp(-x/2)x^{d/2-1}$.

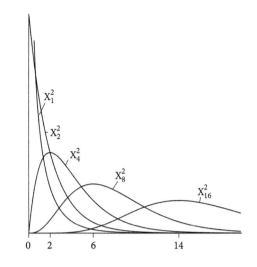

Figure 5.15 Examples of chi-squared distributions. The suffices give the degrees of freedom.

The distribution has mean d and variance $2d$. A random variable with a χ^2_d-distribution occurs as the sum of d independent squared standard normal random variables. Since random variations can usually be assumed to be observations from normal distributions, the chi-squared distribution is often used when assessing the appropriateness of a model as a description of a set of data.

6
Models, p-values, and hypotheses

6.1 Models

There are two sorts of models: those that we propose in advance of seeing the data, and those that we devise having seen the data. In this chapter we focus on the former.

As an example of a model formulated before seeing the data, consider the rolling of a six-sided die. The obvious model is that the die is fair, with each of the six sides being equally likely to appear. The model will not be entirely correct because there are sure to be imperfections. Maybe ⚀ is very slightly more likely to face downwards than ⚅? Maybe one side is dirtier than another? Maybe there is a scratch? However, unless there is some cheating going on, the bias will probably be so small that the die would have been worn away before we could detect it! The model is not 100% correct, but it should provide a very useful description.

Now suppose that after rolling the die 100 times, we obtain roughly equal counts of the occurrence of ⚀ and ⚅ but we never see any of the other four sides. This result would lead us to strongly doubt that our model was correct. Why? Because we could easily calculate that, if the assumption of fairness had been correct, then these observed data would be highly unlikely. It is this kind of test which we consider in this chapter, the so-called Hypothesis Test.

Example 6.1

A simple model, such as one that specifies that oak trees are at random locations in a wood, has at least two uses. Taken together with a count of the oak trees in sampled equal-sized regions of the wood, the model provides the basis for an initial estimate of the number of oak trees in the entire wood. Examination of the variation in the counts can then be used as potential evidence against the simple model. For example, differences in the terrain or other relevant features, might lead to variations in where the oak trees are likely to be found and hence to a non-random pattern.

Data Analysis: A Gentle Introduction for Future Data Scientists. Graham Upton and Dan Brawn, Oxford University Press.
© Graham Upton and Dan Brawn (2023). DOI: 10.1093/oso/9780192885777.003.0006

6.2 *p*-values and the null hypothesis

According to *Oxford Languages* a hypothesis is 'a supposition or proposed explanation made on the basis of limited evidence as a starting point for further investigation.' This is precisely what the word means to statisticians, who elaborate on the idea by introducing a particular type of hypothesis that they call the **null hypothesis**.

The null hypothesis specifies that the population possesses some statistical feature. Some examples are:

> The mean is 500 g.
> The probability of a rolling six is 1/6.
> Oak trees occur at random.

If we imagined that this was a court of law, then the null hypothesis would be that the prisoner was innocent and the sample data would be the evidence. The prisoner will be given the benefit of the doubt (the null hypothesis is accepted) unless there is evidence of guilt (the observed data are too unlikely for the null hypothesis to be believed).

Consider a hypothesis about the population mean. The evidence of guilt (or otherwise) is provided by the sample mean. If it is very close to the hypothesized value then we would be happy to accept that hypothesis. But the greater the difference between the sample mean and the population mean, the stronger the evidence that the null hypothesis was incorrect. The *p*-value is a measure of the strength of that evidence.

> The *p*-value is the probability that, if the null hypothesis were correct, then the observed result, or a more extreme one, could have occurred.

6.2.1 Two-sided or one-sided?

Usually the question we are asking is 'Do we have significant evidence that this statement is untrue?'. Values that are much greater than the value specified by the null hypothesis would lead to rejection of the hypothesis and values that are much smaller would do likewise. That is the two-sided case.

Suppose, however, that we are examining packaged goods that are supposed to have (at least) some specified weight. If the weight of the goods falls consistently short of the specified weight then we will prosecute, whereas if they are overweight then the customer is fortunate. This is an example of the one-sided case with the null hypothesis being that the mean weight is as specified.

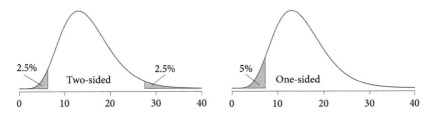

Figure 6.1 Two-sided and one-sided examples of cases corresponding to tail probabilities of 5%.

Figure 6.1 illustrates two cases where the *p*-value is 0.05. In the two-sided case there is a probability of 0.025 in each tail. In the one-sided case the full 0.05 occurs in one tail only.

6.2.2 Interpreting *p*-values

A rough interpretation of a p-value might be as follows:

p-value	Interpretation
$0.01 < p \leq 0.05$	There is some evidence that contradicts the null hypothesis. Ideally we would like more data to provide confirmation.
$0.001 < p \leq 0.01$	There is strong evidence to reject the null hypothesis.
$p \leq 0.001$	There is overwhelming evidence against the null hypothesis.

To be clear, the case $p \leq 0.001$ is stating that if the null hypothesis were true, and we were to look at a thousand situations similar to that currently studied, then perhaps only one of those would have results as extreme as has occurred in our case.

Example 6.2

As an example, suppose we test the hypothesis that a population has mean 5, using a random sample of 100 observations. Figure 6.2 shows a histogram of the sample data (which was a random sample from a normal distribution with mean 6 and standard deviation 2).

The sample mean is 6.1. To assess whether this is unreasonably far from the hypothesized mean of 5, we use a so-called *t*-**test.** This makes use of the *t*-**distribution**, which is a symmetric distribution very similar to the normal distribution.

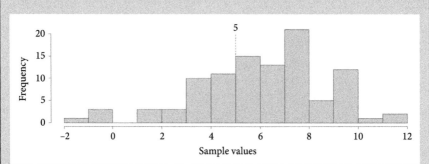

Figure 6.2 Histogram for a sample of 100 observations from a population with mean 6. The hypothesized mean was 5.

The output from R states that the *p*-value is '6.955e-05', which is a shorthand for 0.00006955. In other words, if the mean were 5, then the probability of rolling a sample mean this different, or more different, from the hypothesized value is less than 7 in 100,000. The null hypothesis would most certainly be rejected: we would conclude that the true population mean was around 6.1 (since that was the sample mean).

6.2.3 Comparing *p*-values

Individual *p*-values should always be quoted so as to leave the reader or researcher able to decide how convincing is the evidence against the null hypothesis. This is particularly useful when comparing *p*-values.

When a model involves several explanatory variables[1] then some of these variables will inevitably be more important than others. Typical computer output will associate *p*-values with each variable. These *p*-values are quantifying the importance of each variable by reporting how reasonable it would be to remove that variable from the model. As we will see in later chapters, removal of a variable is equivalent to accepting the null hypothesis that a model parameter has a zero value.

---

[1] Knowledge of the values of explanatory variables, which are also known as background or predictor variables, may help to explain the variation in the values of the variable of prime interest.

6.2.4 Link with confidence interval

Consider, as an example, a 95% confidence interval for the population mean, μ, based on the sample mean, $\bar{x}$. The interval is centred on $\bar{x}$ and takes the form:

$$(\bar{x} - k) \qquad \text{to} \qquad (\bar{x} + k)$$

for some value of k. If, beforehand, we had hypothesized that μ had a value that turns out to lie in the interval, then we would probably look smug and say 'I thought as much'. We would have, in effect, accepted our null hypothesis.

However, suppose that the value that we anticipated lies outside the 95% confidence interval. In that case it must either be higher than the top of the interval or lower than the bottom of the interval. With 95% inside the interval this implies that it lies in one or other of the two 2.5% tail regions. So, the probability of the observed value, or a more extreme one (either greater or lesser), will be at most 5%. That probability is simply the *p*-value.

Example 6.2 (cont.)

The R output for the *t*-test included a 95% confidence interval for the true population mean as being between 5.565 and 6.598. Suppose that our null hypothesis had been that the population mean was not 5 but 5.6. This time, when we perform a *t*-test we get the information that the *p*-value is 0.067. Any hypothesized value lying inside the 95% confidence interval will lead to a *p*-value greater than 5%.

6.3 *p*-values when comparing two samples

In this section we give two examples of the use of *p*-values. Both situations involve the comparison of a pair of samples.

6.3.1 Do the two samples come from the same population?

A diagram that helps to answer this question is the **quantile-quantile plot**. which is usually referred to as a **Q-Q plot**. We met quantiles in Section 2.2.2.

To see how it works, suppose that there are 500 observations in one data set and 200 in another. For the larger data set, working for convenience with the special case of percentiles,[2] the values of the first five are the 5th, 10th, 15th, 20th, and 25th largest values. For the smaller data set the first five quantiles are the 2nd, 4th, 6th, 8th, and 10th largest values. In the Q-Q plot the quantiles for one data set are plotted against those for the other data set. If the two sets come from the same population, then the plot will be an approximately straight line.

Example 6.3

As an example we use three data sets, *x*, *y*, and *z*. Each has 100 observations. Sets *x* and *y* consist of random samples from one population, whereas *z* is a sample from a different population (but with the same mean and variance as the first population).

Figure 6.3 compares the Q-Q plot for our two samples *x* and *y* with the Q-Q plot for *x* and *z*.

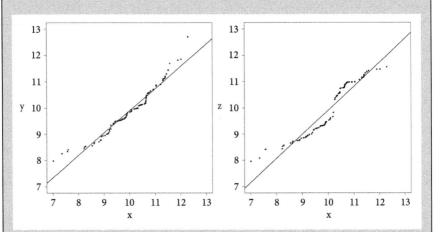

Figure 6.3 Two Q-Q plots: (a) The samples *x* and *y* come from the same distribution. (b) The samples *x* and *z* come from different distributions.

At the ends of a Q-Q plot there is often some discrepancy from the central line as these cases refer to extreme results that are, by definition, rather variable and uncommon. However, the centre of the plot should lie close to the line. For the plot of *x* against *y* there is a degree of wobble, but it appears random, whereas for the plot of *x* against *z* there are marked and coherent divergences from the central line.

---

[2] Quantiles that divide the ordered data into 100 sections.

6.3.1.1 Kolmogorov–Smirnov test

The Q-Q plot does not provide a test, just a visual suggestion. By contrast, the Kolmogorov–Smirnov (KS) test provides a formal appraisal of the null hypothesis that the two samples come from the same distribution. The test statistic is D, the maximum difference between the two cumulative frequency distributions (Section 2.5) of the samples being compared.

Example 6.3 (cont.)

Figure 6.4 (a) compares the cumulative distributions of x and y, which were samples of 100 observations sampled from the same population. There is an appreciable gap between the two curves, but the KS test reports that the maximum difference, 0.18, could reasonably have occurred by chance. The p-value is reported as being 0.078, which is greater than 5%. The null hypothesis that the two samples came from the same population would (correctly) be accepted.

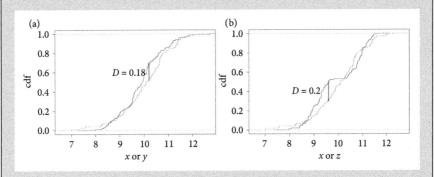

Figure 6.4 Superimposed cumulative distribution functions for (a) x and y, (b) x and z.

Figure 6.4 (b), which compares the cumulative distributions of x and z, is very similar. But here the gap is larger (0.20) and, the KS test reports a p-value of 0.037. Since this is smaller than 5%, there is reason to doubt that the samples have come from the same population (though, since the p-value is not really really small, that decision would be made cautiously).

6.3.2 Do the two populations have the same mean?

The KS test compared the entirety of two samples. If all that is of interest is whether two populations have the same mean, then a *t*-test will do the trick.

Example 6.3 (cont.)

The means of the two populations being compared (x and z) were arranged to be identical (both equal to 10). It is therefore no surprise that the sample means are very similar (10.0 and 9.9). The t-test duly reports a large p-value (0.48) which is much bigger than 5% and indicates that the null hypothesis of equal means is entirely tenable. The 95% confidence interval for the magnitude of the difference between the population means therefore includes zero (the value corresponding to equality).

7
Comparing proportions

7.1 The 2 × 2 table

Suppose that we have two populations, A and B. Within each population there are two categories of interest which we will call 'success' and 'failure'. We wish to know whether p_A, the probability of a success in population A, is equal to p_B, the corresponding probability in population B. If the two are equal, then the probability of a success is **independent** of the population being sampled.

Within the group of successes there are some that come from population A and some from population B. The same division applies to the failures. If the probability of a success is independent of the population, then the proportion of successes that come from population A will be equal to the proportion of failures that come from that population.

> A test of the equality of two proportions is equivalent to a test of **independence** of the outcome on the classifying variable.

Suppose that the data are as shown in Table 7.1.

Table 7.1 Hypothetical observed cell frequencies and their totals.

	A	B	Total
Success	10	20	30
Failure	30	40	70
Total	40	60	100

If the probability of a success is not affected by whether an observation belongs to A or to B, then (using the Total column) the best estimate of the probability of a success is $30/100 = 0.3$. If success or failure is independent of whether an individual belongs to A or B, then there would be $0.3 \times 40 = 12$ successes for A. Assuming that P(success) is indeed 0.3, the complete outcome would be the cell frequencies given in Table 7.2. The

Data Analysis: A Gentle Introduction for Future Data Scientists. Graham Upton and Dan Brawn, Oxford University Press.
© Graham Upton and Dan Brawn (2023). DOI: 10.1093/oso/9780192885777.003.0007

Table 7.2 As Table 7.1 but showing cell frequencies corresponding to the independence model.

	A	B	Total
Success	12	18	30
Failure	28	42	70
Total	40	60	100

greater the difference between the observed counts (Table 7.1) and the ideal independence outcome (Table 7.2), the less likely is it that independence is a correct description of the situation. The question therefore is how best to compare the two sets of numbers. Using the notion of likelihood (Section 5.1.1), the answer is to use the so-called **likelihood-ratio statistic**,[1] G^2, given by

$$G^2 = 2 \sum_{\text{Every combination}} Obs. \times \ln(Obs./Exp.), \qquad (7.1)$$

where *Obs.* refers to an observed count in Table 7.1 and *Exp.* refers to the ideal values given (in Table 7.2) for the current model (which in this case is the model of independence). Here the *Obs.* values are (10, 20, 30, and 40), while the corresponding *Exp.* values are (12, 18, 28, and 42).

In the situations likely to be encountered by data scientists, where there are plenty of data, the distribution of G^2 can be approximated by a chi-squared distribution (Section 5.6.3). In the case of a table with two rows and two columns (such as Table 7.1), the reference chi-squared distribution has a single degree of freedom.

Example 7.1

According to the 1990 report of the British Board of Trade, when the *Titanic* hit the iceberg in 1912 there were 862 male crew and 805 adult male passengers on board. There were 338 male survivors. A possible question of interest is whether the outcome (died or survived) was independent of a person's role (crew or passenger). The relevant numbers are shown in Table 7.3. A table such as this, which would be described as a **2 x 2 table**, is an example of a **contingency table**.

---

[1] The quantity G^2 results from the comparison of two likelihoods. The likelihood, which was introduced in Section 5.1.1, can be described as the probability of obtaining a future set of data identical to the current data under a given model. Here the current model is the model of independence. We compare that likelihood with the maximum possible likelihood by using the natural logarithm of their ratio.

Table 7.3 The numbers and fates of adult male passengers and male crew on board the *Titanic* at the time of its collision with an iceberg in 1912.

	Crew	Passenger	Total
Survived	192	146	338
Died	670	659	1329
Total	862	805	1667

The probability of survival for a male crew member was 192/862=0.22. The probability for an adult male passenger was rather less: 146/805=0.18. Could the difference be due to chance? To assess this we first calculate the values resulting from a a precise fit to the independence model. The values are shown in Table 7.4. The resulting value of G^2 is 4.42 and the question is whether this is such an unusually large value that our model (independence) should be rejected.

Table 7.4 The values obtained when the independence model is applied to the frequencies in Table 7.3.

	Crew	Passenger	Total
Survived	174.8	163.2	338
Died	687.2	641.8	1329
Total	862	805	1667

Comparing the value of G^2 to a chi-squared distribution with one degree of freedom, we find that the probability of that value, or a more extreme one, is about 0.035. Since this is less than 0.05, we would claim that, at the 5% level, there was a significantly higher proportion of surviving male crew than there were surviving male passengers. However, this statement needs to be put into context. Priority was evidently given to women and children, with nearly 70% of the 534 women and children on board surviving; we might speculate that some male passengers sacrificed themselves for their loved ones.

7.2 Some terminology

There are a good many rather specialised terms that are used in connection with 2×2 tables. We now briefly introduce some of them.

7.2.1 Odds, odds ratios, and independence

We are continuing with situations that are classified either as a 'success' or a 'failure'. The **odds** of a success is given by:

$$\text{odds} = \frac{P(\text{outcome is a success})}{P(\text{outcome is a failure})}.$$

The **odds ratio**, is the ratio of the odds for two categories:

$$\text{odds ratio} = \frac{P(\text{a success in category A})}{P(\text{a failure in category A})} \Big/ \frac{P(\text{a success in category B})}{P(\text{a failure in category B})}.$$

If the probability of a success is the same for both categories, the odds ratio is 1 and its logarithm will be zero.[2]

Example 7.1 (cont.)

The odds on survival for male crew was 192/670, while for male passengers it was 146/659. The odds ratio is therefore (192/670)÷(146/659) ≈ 1.3.

7.2.2 Relative risk

Suppose that we are comparing two treatments, A and B, for their success in curing a deadly disease. Suppose that the chance of survival with treatment A is 2%, whereas the chance of survival with treatment B is 1%. We would strongly encourage the use of treatment A, even though the difference in the probabilities of survival is just 1%. A report of this situation would state that treatment A was twice as successful as treatment B. In a medical context, the ratio of the success probabilities, R, is called the **relative risk**.

Notice that the relative risk for failures is different to that for successes. Advertisers of the merits of a treatment will be sure to choose whichever makes the best case for their treatment.

---

[2] Although the log(odds ratio) seems unreasonably obscure in the current context, it is useful in more complex situations.

Example 7.1 (cont.)

For the Titanic situation the relative risk for the male passengers as opposed to crew is 146/805 ÷ 192/862 ≈ 0.81. This implies that the probability of surviving for a male passenger was only about 80% of that for a male crew member, which sounds rather dramatic. If we focus instead on those who died and calculate 659/805 ÷ 670/862, then we find that the risk of dying was just 5% greater for the male passengers.

7.2.3 Sensitivity, specificity, and related quantities

Consider the following 2×2 table:

	Patient has disease	Patient without disease	Total
Patient tests positive	**True positives** (a)	**False positives** (b)	r
Patient tests negative	**False negatives** (c)	**True negatives** (d)	s
Total	m	n	N

The conditional probabilities of particular interest to the clinician are the following:

Sensitivity: P(A patient with the disease is correctly diagnosed)= a/m.
Specificity: P(A patient without the disease is correctly diagnosed)= d/n.

However, a test that suggested that *every* patient had the disease would have sensitivity = 1 (which is good), but specificity = 0 (bad). Both quantities need to be large. A useful summary is provided by **Youden's index:**[3]

$$J = \text{Sensitivity} + \text{Specificity} - 1. \qquad (7.2)$$

While J is a good measure of the usefulness of a test, it gives little information concerning the correctness of the diagnosis.

The conditional probabilities of particular interest to the patient are:

Positive predictive value (PPV): P(Diagnosis of disease is correct)= a/r.
Negative predictive value (NPV): P(Diagnosis of no disease is correct)= d/s.

[3] William John Youden (1900–1971) was an American chemical engineer who became interested in statistics when designing experimental trials.

However, the values of PPV and NPV will be affected by the overall preva-lence (m/N) of the disease. For example, if every patient had the disease then PPV = 1 and NPV = 0. An associated term is the **false discovery rate** b/r (=1 - PPV).

7.3 The $R \times C$ table

We now turn to the more general situation where a table provides informa-tion about two variables, with one variable having R categories and the other having C categories. For example, a table such as Table 7.5:

Table 7.5 Hypothetical observed cell frequencies and their totals.

	A_1	A_2	A_3	A_4	Total
B_1	10	20	30	40	100
B_2	20	40	60	80	200
B_3	30	60	90	120	300
Total	60	120	180	240	600

In this very contrived table, the variable A has four categories, with proba-bilities 0.1, 0.2, 0.3, and 0.4, while the variable B has three categories with probabilities 1/6, 2/6, and 3/6. Furthermore the two variables are com-pletely independent of one another. This can be seen from the fact that the frequencies in each column are in the ratio 1 to 2 to 3.

Real data are not so obliging, but the question of interest remains: 'Are the variables independent of one another?' If they are completely indepen-dent, then knowing the category for one variable will not help us to guess the category of the other variable.

The values that would occur, if the variables were completely independent of one another, are again found by the simple calculation:

$$Exp = \frac{\text{Row total} \times \text{Column total}}{\text{Grand total}}.$$

Thus, for the top left frequency, the Exp value is given by $\frac{100 \times 60}{600}$ = 10, which in this case is equal to the observed value.

A test of the overall independence of the two variables is again provided by G^2, with the reference chi-squared distribution having $(R-1)(C-1)$ degrees of freedom.

Example 7.2

The passengers on the *Titanic* occupied three classes of cabin. We now inquire whether survival of male passengers was independent of their cabin class. The data are shown in Table 7.6.

Table 7.6 The numbers and fates of male passengers on the *Titanic*, with passengers subdivided by cabin class.

	Class 1	Class 2	Class 3	Total
Survived	57	14	75	146
Died	118	154	387	659
Total	175	168	462	805
Approximate proportion surviving	1/3	1/12	1/6	

There appear to be considerable variations in the survival rates. Table 7.7 gives the *Exp* values based on independence of survival and cabin class.

Table 7.7 The average numbers expected in the various category combinations if cabin class and survival were independent.

	Class 1	Class 2	Class 3	Total
Survived	31.7	30.5	83.8	146
Died	143.3	137.5	378.2	659
Total	175	168	462	805

In this case the value of G^2 is 35.2. There are $(2-1)(3-1) = 2$ degrees of freedom for the approximating chi-squared distribution, which therefore has mean 2. If survival were independent of cabin class, then a value of G^2 as large as that observed would only occur on about two occasions in one hundred million. In other words we have overwhelming evidence of a lack of independence.

7.3.1 Residuals

The Pearson[4] residual, r is defined by:

$$r = \frac{Obs - Exp}{\sqrt{Exp}}. \tag{7.3}$$

---

[4] Named after Karl Pearson (1857–1936), an English biometrician who proposed the chi-squared test in 1900.

The idea behind the division by $\sqrt{Exp}$ is to take account of the magnitude of the numbers concerned. For example, 101 differs from 102 by the same amount as 1 differs from 2, but the latter represents a doubling, whereas the former represents a negligible change. The so-called **standardized residual** replaces $\sqrt{Exp}$ by a slightly smaller value. The difference is negligible in large samples.

As a guide, when the independence model is correct, the r-values will be in the range $(-2, 2)$, or, exceptionally, $(-3, 3)$. Values with greater magnitude indicate category combinations where the independence model is certainly failing.

Remember that the values of $(Obs - Exp)$ sum to zero, so negative r-values will be offset by positive r-values.

Example 7.2 (cont.)

The Pearson residuals for the Titanic data are given in Table 7.8, with the most extreme value shown in bold type.

Table 7.8 The Pearson residuals for the model of independence of survival on cabin class.

	Class 1	Class 2	Class 3
Survived	**4.5**	-3.0	-1.0
Died	-2.1	1.4	-.5

The residuals highlight that it is the high proportion of survivors amongst male passengers who occupied Class 1 cabins that results in a failure of the model of independence.

7.3.2 Partitioning

Occasionally it is useful to divide a table into component parts, so as to make clear departures from independence. If there are d degrees of freedom for the G^2 statistic, then notionally it would be possible to present the data as d separate 2 by 2 tables, or some other combination of tables. Whatever subdivision is used, the degrees of freedom for the separate tables will sum to d and the sum of their G^2-values will equal the G^2-value for the overall table.

Example 7.2 (cont.)

The obvious subdivision for the cabin class data is to treat Class 1 cabins as a class apart, as shown in Table 7.9.

Table 7.9 Partition of the class data into two 2 × 2 tables.

	Class 1	Other classes	Total		Class 2	Class 3	Total
Survived	57	89	146		14	75	89
Died	118	541	659		154	387	541
Total	175	630	805		168	462	630

The two G^2-values, each with 1 degree of freedom are 28.2 and 7.0. The latter is also unusually large (tail probability less than 1%) but no Pearson residual has a magnitude greater than 2, suggesting that the evidence is not compelling when it comes to a comparison of the survival rates for the passengers in Classes 2 and 3.

8

Relations between two continuous variables

In previous chapters we have generally looked at one variable at a time. However, the data faced by a data analyst will usually consist of information on many different variables, and the task of the analyst will be to explore the connections between these variables.

In this chapter we look at the case where the data consist of pairs of values and we begin with the situation where both variables are numeric. Here are some examples where the information on both variables is available simultaneously:

x	y
Take-off speed of ski-jumper	Distance jumped
Size of house	Value of house
Depth of soil sample from lake bottom	Amount of water content in sample

Sometimes data are collected on one variable later than the other variable, though the link (the same individual, same plot of land, same family, etc.) remains clear:

x	y
Amount of fertilizer	Amount of growth
Height of father	Height of son when 18

In all of these cases, while the left-hand variable, x, may affect the right-hand variable, y, the reverse cannot be true.

Not all cases follow this pattern however. For example:

x	y
No. of red blood cells in sample of blood	No. of white blood cells in sample
Hand span	Foot length

Data Analysis: A Gentle Introduction for Future Data Scientists. Graham Upton and Dan Brawn, Oxford University Press.
© Graham Upton and Dan Brawn (2023). DOI: 10.1093/oso/9780192885777.003.0008

Whichever situation is relevant, a measure of the strength and type of relation between the variables is provided by the correlation coefficient, with which, having plotted the data, we start the chapter.

8.1 Scatter diagrams

The first step in data analysis is to plot the data in order to get an idea of any relationship and also to check for possible false records. A scatter diagram is simply the plot of the value of one variable against the corresponding value of the other variable.

Example 8.1

In this example we use a data set that gives the numbers of arrests per 100,000 residents for assault, murder, and rape in each of the 50 US states in 1973. Also given is the percentage of the population living in urban areas. Figure 8.1 shows two scatter diagrams involving the incidence of assaults.

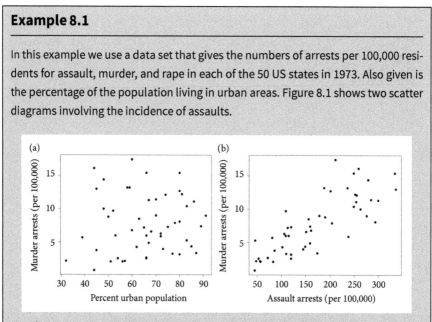

Figure 8.1 Scatter diagrams of the incidence of arrests for murder during 1973, plotted against (a) the percentage of population living in urban areas, and (b) the incidence of arrests for assault. Each data point refers to a single state of the USA.

We might have expected that the crime rates would be greatest in the big cities, but Figure 8.1 (a) shows no obvious pattern. By contrast, Figure 8.1 (b) shows that arrests for murders are common in states where arrests for assault are common. There is a much clearer pattern in Figure 8.1 (b) than in Figure 8.1 (a).

8.2 Correlation

We begin with the measure of the fuzziness in a scatter diagram that Francis Galton[1] called the index of co-relation. These days it is usually denoted by r and referred to simply as correlation. You may also find it described as the **correlation coefficient**, or even as the **product-moment correlation coefficient**.

Correlation can take any value from -1 to 1. In cases where increasing values of one variable, x, are accompanied by generally increasing values of the other variable, y, the variables are said to display **positive correlation**. The opposite situation (**negative correlation**) is one where increasing values of one variable are associated with decreasing values of the other variable. Figure 8.2 shows examples.

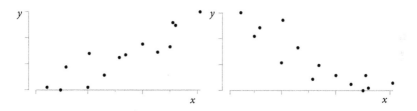

Figure 8.2 Examples of data sets showing positive and negative correlation.

The case $r = 0$ can usually be interpreted as implying that the two variables are unrelated to one another. However, the measure is specifically concerned with whether or not the variables are *linearly* related. Figure 8.3 shows three

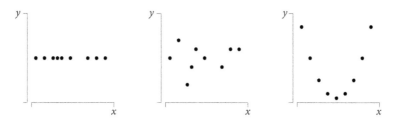

Figure 8.3 Examples of data sets showing zero correlation.

---

[1] Francis Galton (1822–1911), a cousin of Charles Darwin, turned his hand to many activities. He studied medicine at Cambridge. He explored Africa (for which he received the gold medal of the Royal Geographical Society). He devised an early form of weather map introducing the term anticyclone. Possibly inspired by Darwin's work, Galton turned to inheritance and the relationships between the characteristics of successive generations. He proposed the term eugenics as a set of beliefs and practices that aim to improve the genetic quality of a human population. The term co-relation appeared in his 1869 book *Hereditary Genius*.

examples of zero correlation; these include a case where x and y clearly are related, but not linearly. This underlines the need to plot the data whenever possible.

- If the data are collinear (i.e., they lie on a straight line) then (unless that line is parallel to an axis) the correlation is ±1.
- The value of r is unaffected by changes in the units of measurements.
- The correlation between x and y is the same as the correlation between y and x.

Example 8.1 (cont.)

Table 8.1 summarizes the correlations between the four variables in the US states data.

Table 8.1 Correlations between the variables of the US state data. The variables are the percentage of the state population living in urban communities, and the numbers of arrests (per 100,000 residents) for murder, assault, and rape.

	Murder	Assault	% Urban	Rape
Murder	1.00	0.80	0.07	0.56
Assault	0.80	1.00	0.26	0.67
% Urban	0.07	0.26	1.00	0.41
Rape	0.56	0.67	0.41	1.00

The diagonal entries are all 1 because they are looking at the correlation between two identical sets of values. The values above the diagonal are equal to the corresponding entries below the diagonal because cor(x, y) = cor(y, x).

Figure 8.1 (a) showed no obvious pattern and this is reflected in the near-zero correlation (0.07) between the percentage of urban population and arrests for murders. By contrast the pattern evident in Figure 8.1 (b) is reflected in a sizeable positive correlation (0.80) between the relative numbers of arrests for assaults and for murders.

Using R it is possible to look at several scatter diagrams simultaneously. This is illustrated by Figure 8.4. Notice that each pair of variables appears twice, so that, for example, the diagram at top right (the scatter diagram of Murder against Rape) gives the same information as the diagram at bottom left (the scatter diagram of Rape against Murder). The diagrams

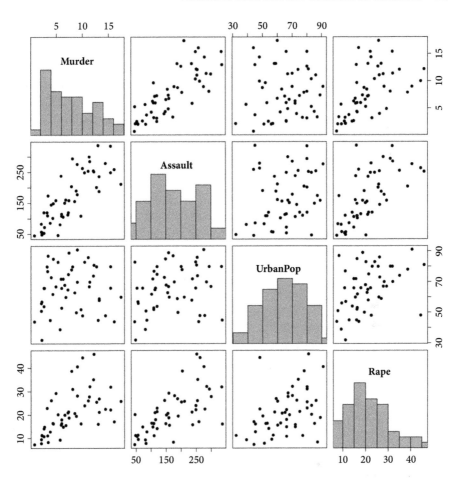

Figure 8.4 A complete set of scatter diagrams for the US arrests.

appear different because the variables are interchanged so that the points are effectively reflected about the 45-degree line.

> ***Correlation does not imply causation.*** *Two variables may appear correlated simply because each is related to some third variable. For example, in children, hand width and foot size are both related to age.*

8.2.1 Testing for independence

If two variables are independent of one another then the population correlation will be zero and the sample correlation should be close to zero. If a 95%

confidence interval for the correlation excludes zero, then we would reject the hypothesis of independence at the 100 − 95 = 5% level.

Before the advent of modern computers the usual test was based on **Fisher's z-transformation**[2] which required various distributional assumptions. Nowadays we can work directly with the observed data by resampling (Section 5.5.1) the *pairs* of values and recalculating the correlation coefficient for each resample. The distribution obtained can then be used to find a bootstrap confidence interval for the correlation coefficient. If the confidence interval includes zero, then the hypothesis of independence would be acceptable.

An alternative test of independence requires resampling the values of one variable only. Suppose that we have n pairs of values for the variables X and Y and that we resample the values for Y[3]. Each resample results in an original X-value being paired with a resampled Y-value so as to produce a new set of n pairs of values. Since the revised pairs use Y-values chosen at random from those available, there will be no connection with the X-values. This means that we are examining the situation where X and Y are uncorrelated, with the aim of seeing what values might occur by chance for the sample correlation.

By resampling only one variable and recalculating the correlations many times we are effectively creating a distribution of feasible correlation values, assuming a true value of zero. In this way we have produced a bootstrap null distribution for correlation. We can compare the original observed correlation value with that distribution and obtain a p-value. If this p-value is substantially less than 0.05 we have clear evidence against the null hypothesis and we can discount the idea of independence.

Example 8.2

We will use data relating to the eruptions by Old Faithful, a geyser in the Yellowstone National Park. The geyser erupts every hour to an hour and a half, but the time intervals vary. It is thought that this variation may be linked to the lengths of the geyser eruptions. The data set consists of 272 pairs of eruption lengths and the waiting times between eruptions, both recorded in minutes. There is strong evidence (a correlation of 0.9) of a link between these two variables. The data are illustrated in Figure 8.5.

---

[2] Sir Ronald Aylmer Fisher (1890–1962) was an English statistician. The 14 editions of his *Statistical Methods for Research Workers* set out the foundations of the subject for twentieth-century statisticians. The test statistic z is given by $z = \frac{1}{2} \ln\left(\frac{1+r}{1-r}\right)$ where ln is the natural logarithm.

[3] It does not matter which variable is resampled.

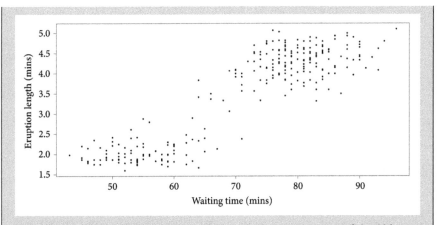

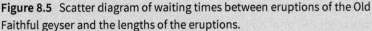

Figure 8.5 Scatter diagram of waiting times between eruptions of the Old Faithful geyser and the lengths of the eruptions.

Using 10,000 resamples on one variable we find correlations in the range $(-0.225, 0.225)$. The observed value of 0.9 is well outside this interval, confirming that there is an undoubted connection between the two variables: the p-value is evidently much less than 1 in 10,000. The 95% confidence interval for the correlation coefficient is found to be $(0.88, 0.92)$.

If the data consist of the entire population then the observed x-values are the only values that can occur, with the same being true for the y-values. We want to know whether the y-values in some way match the x-values.

A subtly different form of resampling is required. We keep the values of one variable unchanged, but randomly *reorder* the values of the second variable. For each reordering we calculate the correlation. If the original correlation is unusually high (or low) compared to those from the random permutations, then we have significant evidence that the variables are related. This is an example of a **permutation test**.

Example 8.1 (cont.)

Figure 8.1 (a) showed no obvious relationship between the numbers of murders (per 100,000 residents) and the percentage of the population living in urban communities. The central 9,500 correlations resulting from 10,000 random permutations of the urban values range from -0.27 to 0.27. This interval easily includes the observed value of 0.07, so the conclusion is that there is no significant interdependence of murder rate and percentage of population living in urban communities for the 50 US states.

By contrast, for the relation between murders and assaults illustrated in Figure 8.1 (b), the observed correlation of 0.80 exceeds all the correlations obtained from 10,000 random permutations of the assault rate. There is an undoubted connection between the two variables.

8.3 The equation of a line

Correlation is a measure of the extent to which two variables are linearly related. If it appears that they are linearly related, then the next question will be 'What is that relation?'. Since the equation of a straight line has the form:

$$y = \alpha + \beta x,$$

our task becomes one of finding appropriate values for α and β.

The constant α is called the **intercept**, and the constant β is the **slope** or **gradient**. Figure 8.6 illustrates the relation between these quantities: when x is zero, y is equal to α, while a unit change in the value of x causes a change of β in the value of y.

8.4 The method of least squares

Correlation treated x and y in the same way, but we now adopt an asymmetric approach. We assume that the value observed for one variable depends upon

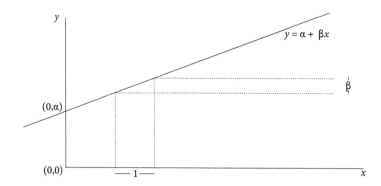

Figure 8.6 Interpretation of the equation for a straight line.

the value taken by the other variable. We will refer to x as the value of the **explanatory variable**, with y being the value of the **dependent variable** (also called the **response variable**).

Suppose that the ith observation has co-ordinates (x_i, y_i). Corresponding to the value x_i, the line will have a y-value of $\alpha + \beta x_i$. The discrepancy between this value and y_i is called the **residual** (see Figure 8.7) and is denoted by r_i:

$$r_i = y_i - (\alpha + \beta x_i). \tag{8.1}$$

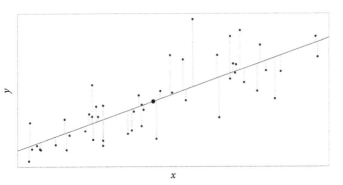

Figure 8.7 The residuals for the line of best fit of y on x are shown as dotted lines. The line goes through $(\bar{x}, \bar{y})$, shown as a larger dot.

If the values of α and β are well chosen, then all of the residuals r_1, r_2, ..., r_n will be relatively small in magnitude. In view of the fact that some of the residuals will be negative and some will be positive, we work with their squares.

> The values chosen for α and β are the values that minimize $\sum r_i^2$.

Example 8.3

A factory uses steam to keep its radiators hot. Records are kept of y, the monthly consumption of steam for heating purposes (measured in lbs) and of x, the average monthly temperature (in degrees C). The results were as follows:

x	1.8	−1.3	−0.9	14.9	16.3	21.8	23.6	24.8
y	11.0	11.1	12.5	8.4	9.3	8.7	6.4	8.5
x	21.5	14.2	8.0	−1.7	−2.2	3.9	8.2	9.2
y	7.8	9.1	8.2	12.2	11.9	9.6	10.9	9.6

We begin by plotting the data (Figure 8.8).

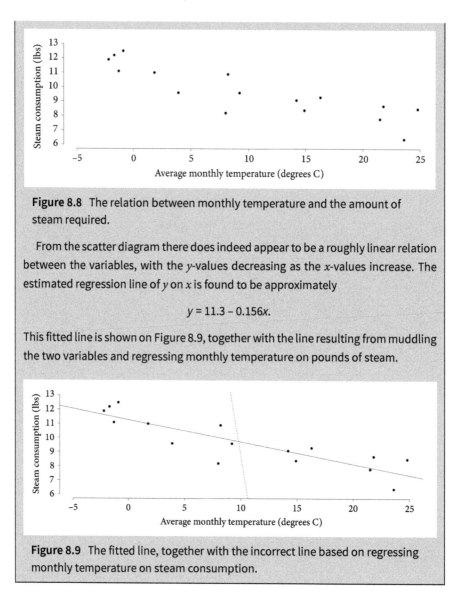

Figure 8.8 The relation between monthly temperature and the amount of steam required.

From the scatter diagram there does indeed appear to be a roughly linear relation between the variables, with the y-values decreasing as the x-values increase. The estimated regression line of y on x is found to be approximately

$$y = 11.3 - 0.156x.$$

This fitted line is shown on Figure 8.9, together with the line resulting from muddling the two variables and regressing monthly temperature on pounds of steam.

Figure 8.9 The fitted line, together with the incorrect line based on regressing monthly temperature on steam consumption.

Always plot the data to make sure that:

1. *Fitting a straight line is appropriate.*
2. *There are no outliers (possibly incorrect records).*
3. *The fitted line goes through the data. If it doesn't then you have probably made an error!*

When using the least squares regression line there is the implicit assumption that every deviation from the fitted line can be considered to be an observation from the same (unknown) normal distribution. This is usually a reasonable assumption, but there are situations where it is clearly incorrect. For example, if we were asked to count the ducks on the village pond, then we would not expect to miscount by very many. However, with the same task at a large reservoir that sort of accuracy would be infeasible.

> *Check that the extent of the scatter about the line is the same for small values of x as it is for large values. If this is not the case, experiment by taking logarithms (or some other simple function) of one or both of the variables.*

8.5 A random dependent variable, Y

In the previous section no mention was made of probability, probability distributions, or random variables. The method of least squares was simply a procedure for determining sensible values for α and β, the parameters of a line that partially summarized the connection between y and x. We now introduce randomness.

If there is an underlying linear relationship then this relationship connects x not to an individual y-value but to the mean of the y-values *for that particular value of x*. The mean of Y, given the particular value x, is denoted by

$$E(Y|x),$$

which is more formally called the **conditional expectation** of Y. Thus the **linear regression model** is properly written as:

$$E(Y|x) = \alpha + \beta x. \tag{8.2}$$

An equivalent presentation is:

$$y_i = \alpha + \beta x_i + \epsilon_i, \tag{8.3}$$

where ϵ_i, the error term, is an observation from a distribution with mean 0 and variance σ^2. Comparing these two equations demonstrates that the regression line is providing an estimate of the mean (the expectation) of Y for any given value of x.

8.5.1 Estimation of σ^2

Assuming that our model is correct, the information about σ^2 is provided by the residuals and an appropriate estimate of σ^2 is provided by:

$$\widehat{\sigma^2} = D/(n-2), \tag{8.4}$$

where n is the number of observations, and D is the **residual sum of squares**:

$$D = \sum_{i=1}^{n} r_i^2. \tag{8.5}$$

8.5.2 Confidence interval for the regression line

For each observed x_i-value, there is a fitted $\widehat{y}_i$ value given by:

$$\widehat{y}_i = \widehat{\alpha} + \widehat{\beta} x_i, \tag{8.6}$$

where $\widehat{\alpha}$ and $\widehat{\beta}$ are the least squares estimates of the unknown regression parameters and $\widehat{y}_i$ provides an estimate of the average of Y when the value of x is x_i.

The confidence interval surrounding a $\widehat{y}_i$ value gives an indication of how accurately the average value of y is known for that particular x-value. The variance associated with $\widehat{y}_i$ is a function of $(x_i - \bar{x})^2$ which implies that the variance is least when $x = \bar{x}$ and steadily increases as x deviates from $\bar{x}$.

8.5.3 Prediction interval for future values

Using Equation (8.3), a future observation, to be taken at x_f, would be given by:

$$\alpha + \beta x_f + \epsilon_f.$$

However, not only do we not know the precise values for α and β but we also cannot know what size random error (ϵ_f) will occur. The uncertainty concerning α and β is responsible for the confidence interval for the regression line. Here, however, we have a third source of variation, so the prediction interval is always wider than the confidence interval for the regression line.

Example 8.3 (cont.)

In Figure 8.10 the inner pair of bounds are the 95% confidence bounds for the fitted regression line. Any line that can be fitted between the bounds would be considered to be a plausible description of the dependence of steam consumption on average monthly temperature.

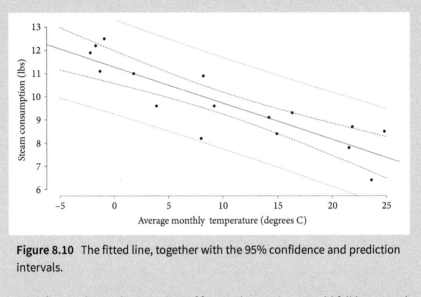

Figure 8.10   The fitted line, together with the 95% confidence and prediction intervals.

According to the predictions, 95% of future observations would fall between the outer bounds. Reassuringly, all the observations actually made do fall within these bounds.

8.6 Departures from linearity

8.6.1 Transformations

Not all relationships are linear! However, there are quite a few non-linear relations which can be turned into the linear form. Here are some examples:

$$y = \gamma x^{\beta} \qquad \text{Take logarithms} \qquad \log(y) = \log(\gamma) + \beta\log(x)$$
$$y = \gamma e^{\beta x} \qquad \text{Take natural logarithms} \quad \ln(y) = \ln(\gamma) + \beta x$$
$$y = (\alpha + \beta x)^{k} \quad \text{Take } k\text{th root} \qquad y^{1/k} = \alpha + \beta x$$

In each case we can find a way of linearising the relationship so that the previously derived formulae can be used. The advantage of linearization is that the parameter estimates can be obtained using standard formulae. Remember that it will be necessary to translate the line back to the original non-linear form of the relationship when the results are reported.

Example 8.4

Anyone owning a car will be dismayed at the rate at which it loses value. A reasonable model states that a car loses a constant proportion of its previous value with every passing year. Thus, if initially a car is valued at V_0 and its value after a year is given by $V_1 = \gamma V_0$, then its value after two years will be $V_2 = \gamma V_1 = \gamma^2 V_0$, and so forth. Thus, after x years, the car will have a value $\gamma^x V_0$. If we take logarithms (to any base) then we have, for example, the linear model:

$$\ln(V_x) = \ln(V_0) + \ln(\gamma)x.$$

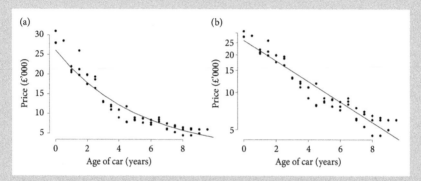

Figure 8.11 (a) Price plotted against age, with the fitted curve derived from (b). (b) As (a), but with price plotted using a logarithmic scale.

Figure 8.11 refers to the Nissan Leaf car. It ignores model revisions and simply reports the list prices of second-hand Leafs and their approximate ages based on their number plates. Figure 8.11 (a) illustrates the data in the usual fashion, whereas Figure 8.11 (b) uses a logarithmic scale on the y-axis. This diagram shows the linear relation between ln(price) and age. The relation is also plotted on Figure 8.11 (a). The value of γ is estimated as 0.82: the annual loss of value was nearly 20% of the previous year's value.

8.6.2 Extrapolation

For many relations no transformation is needed because over the restricted range of the data the relation does appear to be linear. As an example, consider the following fictitious data:

x	y
Amount of fertilizer per m^2	Yield of tomatoes per plant
10 g	1.4 kg
20 g	1.6 kg
30 g	1.8 kg

In this tiny data set there is an exact linear relation between the yield and the amount of fertilizer, namely $y = 0.02x + 1.2$. How we use that relation will vary with the situation. Here are some examples:

1. We can reasonably guess that, for example, if we had applied 25 g of fertilizer then we would have got a yield of about 1.7 kg. This is a sensible guess, because 25 g is a value similar to those in the original data.
2. We can expect that 35 g of fertilizer would give a yield of about 1.9 kg. This is reasonable because the original data involved a range from 10 g to 30 g of fertilizer and 35 g is only a relatively small increase beyond the end of that range.
3. We can expect that 60 g of fertilizer might lead to a yield in excess of 2 kg, as predicted by the formula. However, this is little more than a guess, since the original range of investigation (10 g to 30 g) is very different from the 60 g that we are now considering.
4. If we use 600 g of fertilizer then the formula predicts over 13 kg of tomatoes. This is obviously nonsense! In practice the yield would probably be zero because the poor plants would be smothered in fertilizer!

    Our linear relation cannot possibly hold for *all* values of the variables, however well it appears to describe the relation in the given data.

> The least squares regression line is <u>not</u> a substitute for common sense!

8.6.3 Outliers

An **outlier** is an observation that has values that are very different from the values possessed by the rest of the data. An outlier can seriously affect the parameter estimates. Figure 8.12 shows two examples. In each case all the points should be lying close to the regression line. However, in each case, one y-value has been increased by 30.

In Figure 8.12 (a), there is little change to the slope, but the intercept with the y-axis is slightly altered. The difference appears trivial but is very apparent

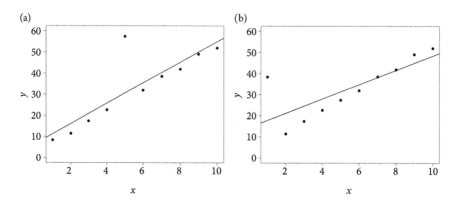

Figure 8.12 (a) An inflated *y*-value for a central *x*-value shifts the fitted line upwards. (b) An inflated *y*-value for an extreme *x*-value drastically alters the slope.

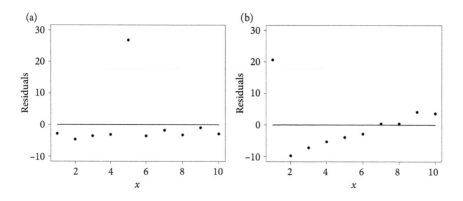

Figure 8.13 The residuals corresponding to the cases illustrated in Figure 8.12.

when the residuals are studied (Figure 8.13): nine of the ten residuals are negative.

In Figure 8.12 (b), because the outlier has an extreme *x*-value, both the slope and the intercept are affected. Notice how all the residuals are affected, not just those for the affected value. There is an obvious pattern including a succession of five negative residuals.

The most common typographical error involves the interchange of neighbouring digits.

For any model, the good data scientist will always look for patterns in the residuals. If a pattern is found then that indicates either an outlier or an incorrect model.

8.7 Distinguishing x and Y

Below are pairs of examples of x and Y-variables. In each case the x-variable has a non-random value set by the person carrying out the investigation, while the Y-variable has an unpredictable (random) value.

x	Y
Length of chemical reaction (mins)	Amount of compound produced (g)
Amount of chemical compound (g)	Time taken to produce this amount (mins)
An interval of time (hrs)	Number of cars passing during this interval
Number of cars passing junction	Time taken for these cars to pass (hrs)

To decide which variable is x and which is Y evidently requires some knowledge of how and why the data were collected.

8.8 Why 'regression'?

Galton (see footnote in Section 8.2) studied inheritance by examining the heights of successive generations of people. He recorded his data in a notebook that is preserved at University College, London. Photographs of the pages can be viewed online;[4] the start of the notebook is shown in Figure 8.14.

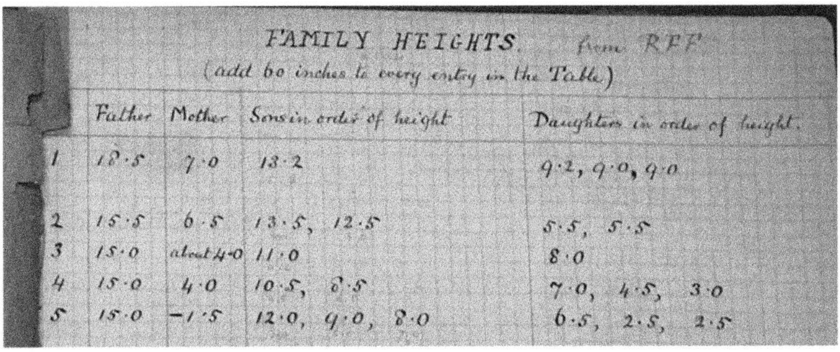
FAMILY HEIGHTS. from RFF
(add 60 inches to every entry in the Table)

	Father	Mother	Sons in order of height	Daughters in order of height.
1	18·5	7·0	13·2	9·2, 9·0, 9·0
2	15·5	6·5	13·5, 12·5	5·5, 5·5
3	15·0	about 4·0	11·0	8·0
4	15·0	4·0	10·5, 8·5	7·0, 4·5, 3·0
5	15·0	−1·5	12·0, 9·0, 8·0	6·5, 2·5, 2·5

Figure 8.14 The start of Galton's notebook recording the heights of family members. Notice the record of the first son's height is given as 13.2, rather than 13.25.

[4] Currently at `http://www.medicine.mcgill.ca/epidemiology/hanley/galton/notebook/index.html`.

Examining Figure 8.14, it is immediately apparent that Galton was not being over-precise with his measurements, since a majority of the heights recorded are given in round numbers of inches. An electronic version of the data is available. Using that data, the 'round-number' bias, is summarized in Table 8.2 using the complete data for fathers and sons.

Table 8.2 The decimal places recorded by Galton show a distinct 'round-number' bias.

Decimal part of height	0	1	2	3	4	5	6	7	8	9
Frequency	676	0	26	4	0	188	0	36	0	0

It seems probable that '0.2' and the occasional '0.3' were Galton's renderings of 0.25, with 0.75 being recorded as '0.7'.

Figure 8.15 plots Galton's data. The left-hand diagram is very clear, but it is misleading, since, because of Galton's rounding of the data, a plotted point may correspond to more than one father-son combination. For example, there were 14 cases where both father and son were recorded as having heights of 70 inches. The right-hand diagram presents the data with each original pair of values being adjusted by the addition or subtraction of small randomly chosen increments. The process is called adding **jitter**.

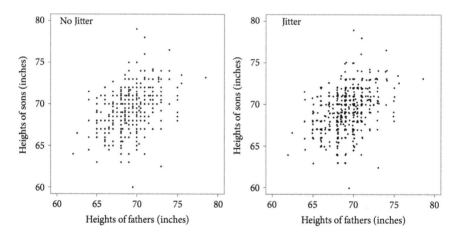

Figure 8.15 Two plots of father's heights against son's heights. The left-hand plot shows the height combinations that were recorded. By adding jitter (see text) to the original data, the right-hand plot gives a slightly clearer idea of the amount of data recorded.

When Galton studied the data, he noticed two things:

1. On average, the heights of adult children of tall parents were greater than the heights of adult children of short parents: the averages appeared to be (more or less) *linearly* related.
2. On average, the children of tall parents are shorter than their parents, whereas the children of short parents are taller than their parents: the values *regress* towards the mean.

These findings led Galton (in a talk entitled 'Regression towards mediocrity in hereditary stature' given to the British Association for the Advancement of Science in 1885) to refer to his summary line drawn through the data as being a **regression line**, and this name is now used to describe quite general relationships.

To visualize Galton's second finding, Figure 8.16 plots the difference between the son's height and the father's height against the father's height.

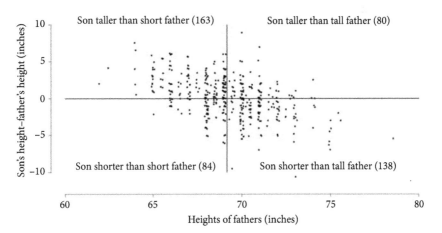

Figure 8.16 On average, short fathers have taller sons; tall fathers have shorter sons.

9
Several explanatory variables

The previous chapter concerned the dependence of one variable (the response variable) on a single explanatory variable. However, there are often several possible explanatory variables. The challenge for the data analyst is to determine which are the key variables and which can safely be ignored. This process is often called **feature selection**.

The **saturated model** is the most complicated model that can be considered for the given data. Although it will be the best fitting model, it will also employ the most parameters, and may therefore be too complicated to be useful. In effect the saturated model is attempting to describe not only the underlying trend (a good thing) but also the random variations present in the sample (a bad thing). This is described as **overfitting**. Such a model can provide less reliable predictions than those from a model that picks out only the underlying trend. We will return to this point in Section 9.5.

A measure of the lack fit of a model is provided by the **deviance** which compares the likelihood (Section 5.1.1) for the model with that for the saturated model.

Example 9.1

Suppose we toss a coin 100 times and obtain 43 heads and 57 tails. The saturated model (echoing the observed outcome) would propose that the probability of a head was 0.43 and the probability of a tail was 0.57. For this model, the likelihood, L_s would be given by:

$$L_s = c \times 0.43^{43} \times 0.57^{57}.$$

where the constant c is a count of the number of ways that 43 heads could arise during 100 tosses.

A simpler (and potentially more useful) model would state that the coin is fair: P(head) = P(Tail) = 0.5. The likelihood, L_f, for this model would be given by:

$$L_f = c \times 0.5^{100}.$$

Data Analysis: A Gentle Introduction for Future Data Scientists. Graham Upton and Dan Brawn, Oxford University Press.
© Graham Upton and Dan Brawn (2023). DOI: 10.1093/oso/9780192885777.003.0009

> The deviance for the fair model is given by:
>
> $$D_f = -2\ln(L_f/L_s),$$
>
> where ln signifies the natural logarithm. The value of c is irrelevant as it cancels out when the ratio of the likelihoods is evaluated.

In the next section we see how we may decide between competing models, in a way that takes account of their fit to the data and their complexity.

9.1 *AIC* and related measures

William of Ockham was a Franciscan friar believed to have been born in Ockham (in Surrey, England) in about 1287. He suggested that when there are alternative explanations of some phenomenon, it is the simpler explanation that should be preferred: this is the principle now called **Ockham's razor** or the **principle of parsimony**.

Suppose we have a complicated model, M_c with many explanatory variables, for which the deviance is equal to D_c. There is also the simpler model M_b which involves just a few of the variables that were in M_c, and nothing extra. It is impossible for this simpler model to give a better fit, so it will have a larger deviance, D_b. The question is whether the gain in simplicity offsets the loss of fit.

Example 9.2

As an analogy, suppose that we have three piles of bricks, as follows:

Number of bricks	8	12	20
Weight (kg)	22	38	60

A perfectly accurate description would be: 'There is a pile of 8 bricks weighing 22 kg, a pile of 12 bricks weighing 38 kg, and a pile of 20 bricks weighing 60 kg.' But the description 'There are piles containing 8, 12, and 20 bricks with an average brick weighing 3 kg.' will be much more useful, even though it is not a perfect description.

A useful measure is therefore one that balances model complexity and goodness-of-fit. The most commonly used measure is the **Akaike**[1]

---

[1] Hirotugu Akaike (1927–2009) graduated from the University of Tokyo in 1952. His career was spent in the Institute of Statistical Mathematics where he was Director General between 1986 and 1994. He was awarded the Kyoto Prize (Japan's highest award for global achievement) in 2006.

Information Criterion (*AIC*) which is a function of the deviance and the number of parameters in the model that require estimation.

Since Akaike introduced his measure, a small correction has been suggested. The adjusted measure is referred to as *AICc*.

Very similar in spirit to *AIC* is the **Bayesian Information Criterion** (*BIC*), also known as the **Schwarz criterion**. As with *AIC*, small values are preferable to large ones. However, *AIC* and *BIC* have subtly different motivations: *AIC* seeks to select that model, from those available, that most closely resembles the true model (which will be governed by myriad unmeasured considerations and will rarely be amongst those considered), whereas *BIC* assumes that the correct model *is* amongst those on offer and seeks to identify that optimal model.

With complicated situations that require a data scientist, the true explanation for the data will involve variables unavailable to the investigator. For this reason we recommend using AICc (or AIC) for model selection.

No automated model selection procedure should be used as a substitute for thought. Different procedures may arrive at different conclusions; it is up to the user to make the final judgement.

9.2 Multiple regression

The previous chapter concentrated on the situations where the value exhibited by a response variable might depend upon the value of a single explanatory variable. Often, however, there are several possible explanatory variables. Rather than considering a series of regression models linking the response variable, in turn, to each of the explanatory variables, we can incorporate information about all the explanatory variables in a single model. This is called a multiple regression model. We start with the simple case where there are just two explanatory variables.

9.2.1 Two variables

Suppose we have two explanatory variables, X_1 and X_2, and the continuous response variable Y. The straightforward extension of the linear regression model is.

$$y = \alpha + \beta_1 x_1 + \beta_2 x_2. \tag{9.1}$$

A sensible analysis might begin with a plot of y against each x-variable in turn, possibly with calculation of the correlation in each case. Assuming that the data are acceptable (no obvious errors or outlying values), the AIC values for the separate linear regression models would be compared with that for the multiple regression model to determine which provides the best fit (i.e., has the smallest AIC value).

The data in the example that follows were provided as the `birthwt` file within the MASS library using R.

Example 9.3

As an example we use data collected during 1986 concerning the weights of 189 babies. The data were collected at Baystate Medical Center, Springfield, Massachusetts during 1986. We will return to this data set throughout the chapter, whilst comparing possible models.

For this first example we use the information concerning the mother's age and weight (in lbs). We begin with separate plots of the weights of the babies against mother's weight and mother's age and calculation of the two correlations. The results are shown in Figure 9.1.

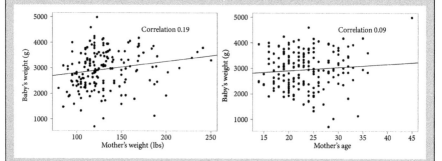

Figure 9.1 Scatter diagrams and fitted regression lines for the dependence of a baby's weight on the mother's weight and age.

There is a plausible positive correlation between the weights of the babies and their mothers. The correlation with the mother's age is much less (0.09) and hawk-eyed readers will have noticed that one mother (with a heavy baby) was 9 years older than the next oldest. When the data for that mother and child are removed, the correlation falls to 0.03.

Continuing by fitting models to the complete data set we get the results summarized in Table 9.1.

Table 9.1 Fits of linear regression models and the multiple regression model to the baby weight data. Here y denotes the baby's weight, x_1 the mother's age, and x_2, the mother's weight.

	α	βMother's age	βMother's weight	AIC
M_0	$y = 2945$			3031
M_{age}	$y = 2656$	$+12.43x_1$		3032
M_{weight}	$y = 2370$		$+4.43x_2$	3026
$M_{age+weight}$	$y = 2214$	$+8.09x_1$	$+4.18x_2$	3028

The model M_0 is the model that estimates the same weight for every baby and ignores the possible influences of any explanatory variables. It is included for completeness and provides a means of judging the usefulness of the explanatory variables. Note that the model M_{age} actually has a higher *AIC* value, suggesting that age (when on its own) should be ignored.

Using *AIC* the best model (lowest *AIC* value) is the linear regression model M_{weight}, with a fit indicated in Figure 9.1 (a). Including information about the mother's age is again unhelpful.

Notice that the βs in the multiple regression model have different values from those in the simpler one-variable models.

> *The relevance of a variable depends on which other variables are in the model.*

9.2.2 Collinearity

When there are many variables that may help to explain the variation in the values of a response variable, it will sometimes happen that two (or more) of these variables are highly correlated because of their dependence on some unmeasured background variable (see Figure 9.2).

In the extreme case where the correlation between variables A and B is 1 (or -1), if one variable is included in the model, then adding the second variable adds no extra useful information: no change in G^2, but an increase in the *AIC* value (because of an unnecessary extra parameter being included).

If the correlation between A and B is large and each variable provides information about the value of a response variable, then a useful model will probably include one of the two variables, but not both. However, which of A and B should be included may depend upon which other variables are present. This can complicate stepwise selection (Section 9.2.5).

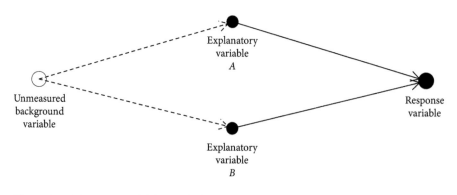

Figure 9.2 When two variables each depend on some (possibly unmeasured) background variable, their values are often correlated.

It is a good idea to examine the correlations between the potential explanatory variables before the main analysis. If two of the variables are highly correlated then consider using just one of them. The extra information gained from including the second variable can be examined after an initial 'final' model has been established.

9.2.3 Using a dummy variable

Categorical variables (e.g., gender or race) can be included as predictors in a regression model by the use of dummy variables. A dummy variable takes the value 0 or 1 depending upon whether a condition is true. Suppose, for example, that we are attempting to estimate a person's weight, w, from knowledge of that person's height, h, using the model:

$$w = \alpha + \beta h.$$

Suppose we collect data from two different countries, A and B, and believe that the linear relation may differ from one country to the other. We will use the dummy variable, D.

A possible model is:

$$w = (\alpha + \gamma D) + \beta h. \tag{9.2}$$

With the dummy variable D taking the value 0 for country A and the value 1 for country B, we are therefore fitting the models:

$$
\begin{array}{ll}
\text{For country A} & \text{For country B} \\
w = \alpha + \beta h & w = (\alpha + \gamma) + \beta h
\end{array}
$$

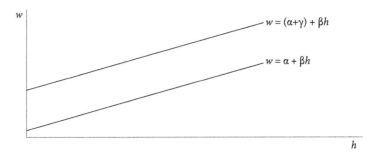

Figure 9.3 The model $w = (\alpha + \gamma D) + \beta h$.

Figure 9.3 shows that, by using the dummy variable with the α term, we are, in effect, fitting parallel lines.

Of course it would also be possible to fit separate models (different slopes and intercepts) by analysing the data from each country separately.

Example 9.3 (cont.)

Amongst other information collected concerning the mothers and their babies was whether the mothers smoked during pregnancy. Figure 9.4 distinguishes between the two groups and shows the regression lines that result when each group are separately analysed.

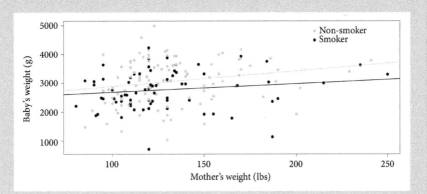

Figure 9.4 The separate regression lines for mothers who smoked during pregnancy and for mothers who did not smoke.

It appears that the non-smoking mothers had slightly heavier babies and, since the two lines are reasonably parallel, this suggests that the model given by Equation (9.2) will be appropriate for the entire data set. The fitted model is:

$$y = (2501 - 272D) + 4.24x,$$

where y and x denote, respectively, the baby's weight and the mother's weight. Here $D = 0$ denotes a non-smoker and $D = 1$ denotes a smoker. These estimates are included in Table 9.2 along with those for simpler models and the *AIC* values.

Table 9.2 Fits of the regression models of baby's weight on mother's weight, with and without the inclusion of the smoking dummy variable. Here y denotes the baby's weight and x the mother's weight.

	α	γSmoker	βMother's weight	AIC
M_0	$y = 2945$			3031
M_{smoker}	$y = 3056$	-284		3026
M_{weight}	$y = 2370$		$+4.43x$	3026
$M_{weight+smoker}$	$y = 2501$	-272	$+4.24x$	3022

The model M_{smoker} states that the mother's weight is irrelevant, with the only determinant for the variation in the weights of the babies being whether or not the mother smoked during pregnancy. The reduction from the *AIC* value for the null model is the same for this model as for M_{weight} (which ignores the smoking factor).

However, the model that includes both, $M_{weight+smoker}$, results in a further reduction in the *AIC* value. This model estimates that a baby will weigh 272 g less for a smoking mother than for a non-smoker of the same weight and also that an increase in the weight of the mother by 1 lb will be matched by an increase of 4.24 g in the baby's weight.

9.2.4 The use of multiple dummy variables

As an example of the use of dummy variables, we supposed, in our introduction to the previous section, that our height and weight data came from two countries, A and B. Suppose instead that the data came from four countries, A to D. We would now need three dummy variables, each of which takes the value 0 or 1 as shown in Table 9.3.

For a model that uses the same slope but different intercepts, we would have:

$$w = \alpha + \gamma_1 D_1 + \gamma_2 D_2 + \gamma_3 D_3 + \beta h,$$

so that, for example, for country C, the model becomes:

$$w = (\alpha + \gamma_2 D_2) + \beta h.$$

Table 9.3 The values for three dummy variables corresponding to four categories of the variable 'country'.

	D_1	D_2	D_3
Country A	0	0	0
Country B	1	0	0
Country C	0	1	0
Country D	0	0	1

Here country A is the reference category with which the other countries are compared.[2]

> If a variable has M non-numeric categories, then we will need M–1 dummy variables.

> Some computer languages apply multiple dummy variables with a single instruction. The language R uses the `factor` command for this purpose.

Example 9.3 (cont.)

Another variable reported for the baby weight data was the mother's race, with the three alternatives being 'White', 'Black', or 'Other'. The result of including this variable is reported in Table 9.4, where the parameter for the intercept is denoted by α, the coefficient of the quantitative variable by β, and the coefficients for the dummy variables by γ.

Table 9.4 The impact of race on the baby's weight.

	α	γ_{Smoker}	γ_{Black}	γ_{Other}	$\beta_{Mum\,wt}$	AIC
M_0	2945					3031
M_{race3}	3103		-383	-297		3025
$M_{race3+weight}$	2486		-452	-241	4.66	3020
$M_{race3+smoker}$	3335	-429	-450	-453		3012
$M_{race3+smoker+weight}$	2799	-400	-504	-395	3.94	3009

Race is evidently an important variable (reducing AIC to 3025). Indeed, each variable reduces AIC. The best fitting model (lowest AIC) includes all three variables.

---

[2] Regression software usually chooses either the first category or the last category as the reference category. An alternative is to compare every category with their average.

We see that mothers described as 'Black' have babies weighing on average about 500 g less than those of mothers classed as 'White' (here chosen arbitrarily as the reference category). The reduction for the mothers described as 'other' is about 400 g. Regardless of race there is a further reduction of 400 g if the mother is a smoker.

The mean weight of a mother is given as 130 lb (just over 9 stones). The average baby weight for a non-smoking White mother is estimated by the model as:

$$2799 + 3.94 \times 130 \approx 3300 \text{ g},$$

whereas that for a Black mother of the same weight who smoked during pregnancy is estimated as:

$$2799 - 400 - 504 + 3.94 \times 130 \approx 2400 \text{ g}.$$

9.2.4.1 Eliminating dummy variables

A variable with m distinct categories will require $(m-1)$ dummy variables. For example, with four countries, we needed three dummy variables. However, we may find that two (or more) categories are sufficiently similar that the same dummy variable can be used for both (or all).

Suppose that Countries C and D are very similar to one another, but are quite different from the other countries. In this case we need only two dummy variables as shown in Table 9.5.

Table 9.5 The values for two dummy variables corresponding to four categories of the variable 'country', when countries C and D are very similar.

	D_1	D_2
Country A	0	0
Country B	1	0
Country C	0	1
Country D	0	1

Example 9.3 (cont.)

Table 9.4 showed that race was an important factor affecting the baby's weight, with both mothers classed as 'Black' and those classed as 'Other' having lighter babies. For some models in the table we see that $y_{\text{Black}} \approx y_{\text{Other}}$. We now consider combining these two categories into a single category which we call 'Non-White'. The results for the competing models are summarized in Table 9.6.

Table 9.6 The impact of race on the baby's weight.

	α	γ_{Smoker}	γ_{Black}	γ_{Other}	$\beta_{Mum\ wt}$	AIC
$M_{race3+smoker+weight}$	2799	−400	−504	−395	3.94	3009
$M_{race2+smoker+weight}$	2849	−412	$\gamma_{Non\text{-}white}$ −431		3.61	3007

Repeating the best model, but with the three categories reduced to two categories results in a further decrease in *AIC*. The simplification is worthwhile.

9.2.5 Model selection

If there are *m* explanatory variables, each of which will be either in or out of the model, then there are 2^m possible models.[3] With *m* = 10, that means there would be 1024 models. With 20 variables there would be more than 1 million possible models. So an automated procedure for choosing the best model is badly needed.

A useful technique is to create the model that contains all of them and require the computer to perform **stepwise selection** starting with that overly complex saturated model. Stepwise selection proceeds by alternately removing (or adding) terms one at a time, with the aim of optimising some function (typically minimizing the *AIC* value).[4]

However, the bottom line is that it is you who is in charge, not the computer! If there is a simple model that explains the vast majority of the variation in the response variable, then the pragmatic choice may be to use that model rather than the computer's more complicated suggestion.

9.2.6 Interactions

It is sometimes the case that the combined effect of two explanatory variables is very different from the simple sum of their separate effects. For example, striking a match while nowhere near a pile of gunpowder makes no noise at all, but the outcome is very different when the match is struck by a pile of gunpowder.

If x_1 and x_2 are two explanatory variables then we can introduce their interaction by adding a product term, $x_1 x_2$, into the model. For example:

---

[3] When interactions (Section 9.2.6) are included the number is much greater still.
[4] Some regression software includes an automated stepwise routine.

$$y = \alpha + \beta_1 x_1 + \beta_2 x_2 + \beta_{12} x_1 x_2.$$

The same procedure works with dummy variables, as the following example illustrates.

Example 9.3 (cont.)

We have found that a mother's race is relevant to the weight of her baby, and that whether she smoked during pregnancy also appears to affect the weight. We now wonder whether the effect of smoking might be more serious for some races than others. Continuing to combine the 'Black' and 'Other' categories, a preliminary investigation involves calculating the mean baby weights for the four race-smoking combinations. The results, which are shown in Table 9.7, certainly suggest that there may be important differences. However, since these results do not take account of the differing mother's weights, a full regression model is required.

Table 9.7 The mean baby weights (and the numbers of mothers) for the four combinations of smoking and race.

	White		Non-white	
Smoker	2827	(52)	2642	(22)
Non-smoker	3429	(44)	2824	(71)
Difference	602		182	

Using the subscripts S, and R, for 'Smoker', and 'Race', we now have the model M_I given by:

$$b = \alpha + \gamma_S x_S + \gamma_R x_R + \gamma_{SR} x_S x_R + \beta m, \tag{9.3}$$

where b and m refer to the weights of babies and mothers, and x_S and x_R are dummy variables. Table 9.8 shows how this works for the four race-smoker combinations (using 0 for 'White' and for 'Non-smoker', and 1 otherwise).

Table 9.8 The outcome of the model for the four combinations of smoking and race.

White non-smoker	$b = \alpha + \beta m$
White smoker	$b = \alpha + \gamma_S + \beta m$
Non-White non-smoker	$b = \alpha + \gamma_R + \beta m$
Non-White smoker	$b = \alpha + \gamma_S + \gamma_R + \gamma_{SR} + \beta m$

The AIC value for the model M_I (Equation (9.3)) for this model is 3006.5, which is marginally less than the value for the model without interactions (3007.3). The difference between these *AIC* values is much less than the other differences recorded between the competing models. This suggests that the interaction (which complicates the interpretation of the model) is of minor importance.

9.2.7 Residuals

Remember that the analysis of a data set does not stop with the choice of a model. Any model is a simplification of reality. We always need to examine the residuals. A residual is the difference between the value observed and the value predicted by the model. Some will be positive, some will be negative, and, hopefully, most will be relatively small in magnitude. A box plot (Section 2.3) is a good way of identifying any unusual values. If there are several similarly unusual values that share some trait not accounted for by the current model, then this points to the need to include an additional explanatory variable, or an interaction, in the current model (or try a completely different model).

Example 9.3 (cont.)

Figure 9.5 shows the boxplot for model M_I Equation (9.3). There are two obvious outlier values. Examination of these cases shows that both mothers had a condition described as 'uterine irritability'. There were 26 other mothers with this condition and their babies were, on average, 600 g lighter than the babies of mothers without that condition. Inclusion of this extra variable in the model reduces the *AIC* value to 2996, which is a considerable improvement on all previous values.

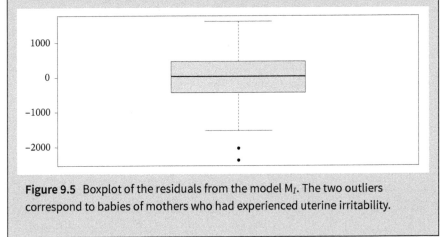

Figure 9.5 Boxplot of the residuals from the model M_I. The two outliers correspond to babies of mothers who had experienced uterine irritability.

9.3 Cross-validation

The principal purposes for fitting a model to a set of data are these:

1. To understand the relations between the variables forming the data set.
2. To summarize the data set in a simple fashion.
3. To predict future values.

The previous emphasis has been on understanding and summarizing the current data. We turn now to considering how useful our present summary model will be for predicting and describing future observations.

The idea of cross-validation is simple. We divide the available data into two parts: a **training set** and a **validation set** (also called a **hold-out set**). The model is estimated (using only the data in the training set) and is then applied to the validation set to hopefully discover that it fits well. The process may be referred to as **out-of-sample validation**.

Suppose that the observed values of the response variable in the validation set are $y_1, y_2, ..., y_n$, with the corresponding estimated values using the model fitted to the training set being $\hat{y}_1, \hat{y}_2, ..., \hat{y}_n$. The key quantity is the **mean squared error** (*MSE*) given by:

$$MSE = \frac{1}{n}\sum_{i=1}^{n}(y_i - \hat{y}_i)^2. \tag{9.4}$$

As an alternative the **root mean squared error** (*RMSE*) may be quoted. This is simply the square root of the mean squared error and has the advantage of being in the same units as the original values. The *MSE* and *RMSE* are effectively estimates of the error variance and standard deviation that will accompany future estimates based on the selected model.

The aims of cross-validation are to estimate the error variance, and to be assured that a reasonable choice of model has been made. With that reassurance, the parameter estimates subsequently used are those based on fitting the selected model to the entire data set. There are several similar approaches.

9.3.1 *k*-fold cross-validation

In this approach the complete data set is divided into k subsets of approximately the same size. Each of these subsets takes on, in turn, the role of the hold-out set, with the remainder in each case being used as the training set. The statistic used in such a case is the average of the k *MSE* calculations over the k choices for the hold-out set.

The choice of value for k is not critical, though the values 5 and 10 are often recommended. However, it may be more useful to choose a smaller value, such as 2 (when each half of the data set is used alternately as the training set and the testing set), or 3 (when there will be a 50% overlap between any pair of training sets). The advantage of these small values of k is that it will be possible to examine the extent to which parameter estimates are affected by the particular data used for training. If there is close agreement in the various parameter estimates, then this will provide reassurance that the model will be appropriate for use with future data.

- *As ever, there will be a need for experimentation. For example, if we are unsure as to which of rival models to use (because of similar AIC values), then a study of their comparative MSE (or RMSE) values for each test set may be useful.*
- *If we find that there are a few observations that are persistently poorly estimated, then this might suggest re-estimating the models with those observations removed, or with the addition of a relevant variable that had previously been omitted.*

Example 9.3 (cont.)

Starting with a model that uses almost all the variables available, together with some interactions, stepwise selection (Section 9.2.5) leads to a model, M_S, that includes the combined 'Non-White' dummy variable. The model includes several other dummy variables. One of these, which indicates whether the mother has experienced urinary infection, is associated with the parameter y_u. Using the entire data set, this model has the AIC value 2988. Replacing the single 'Non-White' dummy variable by the two dummies 'Black' and 'Other' (model M_{S2}) results in a slightly larger AIC value (2990).

Using $k = 3$. we can examine whether M_S has a lower RMSE (or AIC-value) than M_{S2}, using each of the three test sets, and also whether there is much variation in the parameter estimates from one trial set to another. A selection of the results are given in Table 9.9.

Table 9.9 With $k = 3$, The AIC value and the value of the parameter, y_u (see text) are shown for the test sets using three-fold cross-validation. The RMSE values are shown for each model.

Model	First third			Second third			Third third		
	RMSE	AIC	y_u	RMSE	AIC	y_u	RMSE	AIC	y_u
M_S	628	2000	−704	664	1994	−487	682	1992	−521
M_{S2}	630	2002	−709	672	1994	−485	695	1993	−518

Comparing the two models, we can see that the *AIC* value for model for model M_{S2} is never smaller than that for model M_S, confirming that the latter is the preferred model. Similarly the *RMSE* values (over the remaining two-thirds of the data set, for each choice of a third) show that M_S gives slightly more accurate predictions.

However, when we compare the estimates across the three cases, there are considerable variations that are exemplified by the estimates for y_u using model M_S ranging from –487 to –704. The conclusion is that there is a good deal of uncertainty concerning the precise importance of "uterine irritability", though there is no doubt (because of the lower *AIC* value found previously) that this parameter *is* important.

9.3.2 Leave-one-out cross-validation (LOOCV)

In the previous section we advocated a small value of k as a means of exploring the reliability of the chosen model. However, the principal purpose of cross-validation is to estimate the uncertainty when the chosen model is applied to future observations. In this case, with n observations, the choice of $k = n$ is ideal. In this case the trial set consists of all the observations bar one (the test observation). For large n the model based on $n - 1$ of the observations will be well nigh identical to the model that will be used in future (the model based on all n observations). So the comparison of the predicted value based on the $(n - 1)$-model will be virtually identical to that based on all n observations, so the difference between the predicted and observed values (for the nth observation) will be a very accurate representation of a typical error.

Example 9.3 (cont.)

For the model M_S, the LOOCV estimate of *RMSE* (averaged over all the cases) is 657.

9.4 Reconciling bias and variability

Suppose that we have arrived at the model:

$$\hat{y} = \hat{\alpha} + \hat{\beta}_1 x_1 + \hat{\beta}_2 x_2 + \cdots + \hat{\beta}_m x_m,$$

where $\hat{\alpha}, \hat{\beta}_1$, etc., are all the least squares estimates (Section 8.4) of their respective parameters.

Good news: A property of any least squares estimate is that it is unbiased: if we repeat the procedure many times, then the long-term average of the estimate will be precisely correct. That being the case, assuming the model is correct, the long-term average value of $\hat{y}$ will exactly equal the true value.

Bad news: Least squares estimates are only *estimates*. Each new set of data taken from the same population will give a new estimate Yes, *on average* the results are correct, but associated with every $\hat{\beta}$-value there is variability. The more β-values that require estimation, the greater the uncertainty in the estimate of y.

Using the least squares estimates gives an estimate that is, on average, absolutely correct. However, because of the inherent uncertainty we may have been unlucky and the estimate that we have got may be far from correct. It might be better to use instead a procedure that leads to biased estimates that are less variable.

A convenient measure is again the *MSE* (the mean squared error), which is the average value of $(y - \hat{y})^2$. This is equal to the variance of the estimate added to the square of the bias. Figure 9.6 illustrates a case where the unbiased procedure has variance 400 (giving $MSE = 400+0 = 400$), whereas the biased procedure has a variance of 100 and a bias of 5 (giving $MSE = 100+5^2 = 125$).

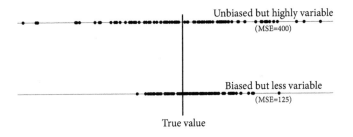

Figure 9.6 A less variable, but biased procedure can have a smaller mean squared error than an unbiased procedure.

9.5 Shrinkage

Shrinkage refers to any procedure that reduces the variability of the model parameter estimates and hence the variability of the predictions.

The standard least squares procedure focuses on minimizing the sum of squared errors, D, given by:

$$D = \sum_{i=1}^{n}(y_i - \hat{y}_i)^2,\qquad(9.5)$$

where y_i is the ith of n observed values and $\hat{y}_i$ is the corresponding estimated value. By contrast, shrinkage estimates are formed by minimizing:

$$D + g(\widehat{\beta_1}, \widehat{\beta_2}, ..., \widehat{\beta_m}), \tag{9.6}$$

where g is some positive function. One choice for the function g is the sum of squares of the coefficients, so that one is minimizing R given by:

$$R = D + \lambda \sum_{i=1}^{m} \widehat{\beta_i}^2. \tag{9.7}$$

This is known as **ridge regression.**

An alternative is to minimize L, given by:

$$L = D + \lambda \sum_{i=1}^{m} |\widehat{\beta_i}|, \tag{9.8}$$

where $|x|$ means the absolute value of x. This is called the **lasso** (which stands for 'least absolute shrinkage and selection operator'). Using the lasso often results in some of the βs being given the value zero, which means that the corresponding explanatory variables are eliminated from the model. The result is a model that is easier to interpret, and gives more precise (but possibly slightly biased) estimates.

In either case, the parameter λ is known as a **tuning constant.** The choice $\lambda = 0$ gives the usual least squares model, with increasing λ placing ever more attention on the β-values. The best value for λ is chosen using cross-validation.

9.5.1 Standardization

Suppose that y is the wear of a car tyre (measured in mm) and the single explanatory variable is x, the distance travelled. If x is measured in thousands of miles, then maybe $\beta = 1$. However, if x is measured in miles, then $\beta = 0.001$. The fitted values will be the same, whichever units are used for x, so that the value of D in Equations (9.7) or (9.8) will be unaltered. The value of $g(\beta)$ will be greatly changed, however.

The solution is standardization. Rather than working with x, we work with the standardized value, x', given by

$$x' = \frac{x - \bar{x}}{s}, \tag{9.9}$$

where $\bar{x}$ is the mean of the x-values and s is their standard deviation. This is applied separately to each x-variable, and also to the y-values. The result is that every standardized variable has mean 0 and variance 1.

Both ridge regression and the lasso work with standardized values. When a group of explanatory variables are highly correlated with one another, ridge regression tends to give each of them similar importance (i.e., roughly equal β values) whereas the lasso will give zero weight to all bar one. Some shrinkage routines include standardization automatically.

Example 9.3 (cont.)

For the baby weight data, the model M_{S2} uses 8 explanatory variables. We assess the advantage of shrinkage by using LOOCV. Results are summarised in Table 9.10.

Table 9.10 The values of the root mean square error ($RMSE$) resulting from using LOOCV. Also given is the number of non-zero parameters in the model.

Model	No. of non-zero parameters	RMSE
Ordinary least squares ($\lambda = 0$)	8	656.7
Ridge regression	8	656.5
Lasso	6	655.1

We see that using ridge regression does slightly reduce the $RMSE$ but not as effectively as the lasso. The latter also simplifies the model by setting two parameters to zero.

9.6 Generalized linear models (GLMs)

In previous sections, the response variable Y was a continuous variable (such as length or weight). Although we did not discuss it, the implicit assumption in our modelling was that the underlying randomness was the same for all y-values and could be described by a normal distribution.

However, linear models can also be used when the response variable has many possible distributions. For example, when the response variable is categorical with two categories (e.g., Yes and No) the model used is a logistic regression model. When a categorical response variable has many categories, a slightly different approach may be used and the model is called a log-linear model. These are briefly described in the next sections.

9.6.1 Logistic regression

We now consider the situation where the quantity of interest is p, the probability of a 'success'. We are investigating whether p is dependent on the values taken by one or more explanatory variables.

However, we cannot work with p directly. To see why, consider the following situation where p depends on a single explanatory variable, x. In the example, samples have been taken at five values of x. The table gives the numbers of successes within each sample:

Value of x	1	2	3	4	5
Sample size (n)	25	40	30	48	50
Number of successes	15	26	21	36	40
Success rate (p)	60%	65%	70%	75%	80%

The simple linear relation:

$$p = 0.55 + 0.05x$$

provides a perfect explanation for the data. So, what is the problem? Well, try putting $x = 10$ into this relation. The result is apparently $p = 1.05$. But probability is limited to the range $(0, 1)$.

Of course, extrapolation can always give foolish results, but we should be using a model that would provide a feasible estimate of p for every value of x. Rather than modelling the variation in p with its limited range of $(0, 1)$, we need to model the variation in some function of p that takes the full $(-\infty, \infty)$ range. The answer is the **logit**:

$$\ln\left(\frac{p}{1-p}\right).$$

The values $0, 0.5$, and 1 for p, correspond to the logit values $-\infty, 0$, and ∞. The relation between p and the logit is illustrated in Figure 9.7.

Note that, for p in the range $(0.1, 0.9)$, the logit is very nearly a straight line. As a consequence, if the relation between p and some variable x appears to be approximately linear in this range, then the same will be true for the logit. Thus, instead of using $p = \alpha + \beta x$, we use:

$$\ln\left(\frac{p}{1-p}\right) = \alpha + \beta x. \tag{9.10}$$

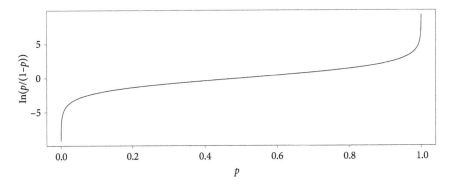

Figure 9.7 The relation between p and the logit is nearly linear for most of the range of p.

with x being the observed value of X. A model of this type is described as a **logistic regression model.**[5]

Example 9.4

We again use the baby birthweight data. Babies weighing less than 2.5 Kg were clas-sified as 'low'. We now examine how the probability of a baby being classified as low (scored as 1), as opposed to normal (scored as 0), is dependent on other variables.

The principal explanatory variable is certainly the mother's weight. In addition we consider two other variables: the number of previous premature labours (0 to 3), and whether or not the mother had a history of hypertension. The three explanatory vari-ables were chosen to illustrate the fact that, as for multiple regression (Section 9.2), any type of explanatory variable can be used (continuous, discrete, or categorical).

Including the number of previous premature labours implies that each premature labour has the same additional effect. An alternative would have been to reduce the number of categories to two: none, or at least one.

The results for a variety of models are given in Table 9.11.

Table 9.11 Some models explaining variations in the logit of the probability of a baby being described as having low weight.

	α	γ_{Hypert}	$\beta_{Premature}$	β_{MumWt}	AIC
M_0	−0.790				237
M_{MumWt}	0.998			−0.0141	233
$M_{Hypert+MumWt}$	1.451	1.855		−0.0187	227
$M_{Labours+MumWt}$	0.641		0.722	−0.0125	229
$M_{Hypert+Labours+MumWt}$	1.093	1.856	0.726	−0.0171	224

---

[5] Any value for x can be converted into a value for p by using $p = \exp(\alpha + \beta x)/\{1 + \exp(\alpha + \beta x)\}$.

Considering only these five models, the lowest *AIC* is given by the most complex. To see how this model works, consider a mother of average weight (130 lb). If she had had no premature labours and no history of hypertension then, the logit would be equal to:

$$1.093 - 0.0171 \times 130 = -1.13$$

which corresponds to a probability of $\exp(-1.13)/(1 + \exp(-1.13)) = 0.244$. By contrast, the logit for a lady of the same weight, suffering with hypertension, and having had two premature labours, would be:

$$1.093 + 1.856 + 2 \times 0.726 - 0.0171 \times 130 = 2.18.$$

This corresponds to a probability of $\exp(2.18)/(1 + \exp(2.18)) = 0.890$, so the model predicts that a mother of average weight with two premature labours and experiencing hypertension has a 90% chance of having a baby of low weight.

9.6.2 Loglinear models

When all the variables are categorical with several categories, the interest in the data may simply be to understand the connections between variables, rather than making predictions (an approach resembling correlation rather than regression). Table 7.5 provided a hypothetical table showing precisely independent variables. In that table each cell count was given by:

$$\text{Cell count} = \frac{\text{Row total} \times \text{Column total}}{\text{Grand total}}.$$

This multiplicative model can be transformed into an equivalent linear model by taking logarithms:

$$\ln(\text{Cell count}) = \ln(\text{Row total}) + \ln(\text{Column total}) - \ln(\text{Grand total}).$$

This simple independence model can be extended to include the effects of several variables and also the interactions between two or more variables as the next example demonstrates.

Example 9.5

The following data are abstracted, with permission of the UK Data Service, from the 2012 Forty-Two Year Follow-Up of the 1970 British Cohort Study (ref. SN7473). Table 9.12 summarizes the answers to questions concerning belief in God and belief in an afterlife. To simplify the presentation, all replies that expressed any doubts have been classed as 'Maybe'.

Table 9.12 Belief in God and belief in an afterlife as expressed by UK 42-year-olds in 2012.

	Male			Female		
	Afterlife	Maybe	No afterlife	Afterlife	Maybe	No afterlife
No God	27	412	747	66	377	248
Maybe	206	1992	266	567	2405	139
God	233	90	23	471	186	24

For simplicity, we will use the suffixes G, A, and S for the variables 'Belief in God', 'Belief in an afterlife', and 'Gender'. In the model definitions we will indicate independence between quantities using a + sign and an interaction using a * sign. We assume that if, for example, the parameter β_{AG} is in the model, then β_A and β_G must also be in the model (in fact, an inevitable consequence of the model fitting procedure).

The simplest model of any interest is the model M_{A+G+S} which states that all three variables are mutually independent.

Table 9.13 Alternative models, their components, and their *AIC* values when applied to the data of Table 9.12.

Model	α	β_A	β_G	β_S	β_{AG}	β_{AS}	β_{GS}	β_{AGS}	AIC value
M_{A+G+S}	✓	✓	✓	✓					3979
M_{A*G+S}	✓	✓	✓	✓	✓				773
M_{A*S+G}	✓	✓	✓	✓		✓			3424
M_{A+G*S}	✓	✓	✓	✓			✓		3693
$M_{A*G+A*S}$	✓	✓	✓	✓	✓	✓			218
$M_{A*G+G*S}$	✓	✓	✓	✓	✓		✓		487
$M_{A*S+G*S}$	✓	✓	✓	✓		✓	✓		3138
$M_{A*G+A*S+G*S}$	✓	✓	✓	✓	✓	✓	✓		191
M_{A*G*S}	✓	✓	✓	✓	✓	✓	✓	✓	166

A glance at the *AIC* values reveals that the models that include the AG interaction should be greatly preferred to the other models: belief in God tends to go hand in hand with belief in an afterlife. Comparing the *AIC* values for the models $M_{A*G+A*S}$ and $M_{A*G+G*S}$, it is apparent that the there is a stronger relation between a person's gender and their belief in the afterlife, as opposed to the connection between their gender and their belief in God.

The model with the lowest *AIC* value is the model M_{A*G*S}. This model states that the association between any two variables is dependent on the third variable. In simple terms, it is the model that says that this is complicated!

The example makes clear that, even with only three variables, there are many models that might be considered. This is not simply a problem with loglinear models, it is true for any modelling that involves many explanatory variables. The solution is to use stepwise selection (Section 9.2.5) together with a careful examination of the results. For example, in the previous example, there was one interaction between variables that was far more important than all the others. In such a case that should be the fact that the data scientist reports in bold type.

10
Classification

The *Cambridge Dictionary* defines classification as 'the act or process of dividing things into groups according to their type'.

The idea is that we have all the background information (B) on an individual but do not know to which of the of classes (C_1, C_2, ...) the individual should be assigned.

In this chapter we introduce some of the simpler procedures that have been used to carry out the classification process.

10.1 Naive Bayes classification

We met Bayes Theorem back in Section 3.5. Expressed in terms of classes (C_1, C_2, ...) and background information (B), the theorem states that:

$$P(C_i|B) = \frac{P(B|C_i) \times P(C_i)}{P(B)}.$$ (10.1)

To decide on the class to which an individual with background B belongs, we first calculate the various products $P(B|C_1) \times P(C_1)$, $P(B|C_i) \times P(C_i)$, ... We then assign the individual to whichever class gives rise to the largest of these products.[1]

If there is information only on one variable, then Equation (10.1) can be used as it stands. Often, however, we have relevant information on m variables, where $m > 1$. This is where the description 'naive' comes in: we assume that these m variables are independent of one another, so that we can replace $P(B|C_i)$ by:

$$P(B_1|C_i) \times P(B_2|C_i) \times \cdots \times P(B_m|C_i),$$ (10.2)

where $B_1, B_2, ..., B_m$ are the m variables.

---

[1] We do not need to calculate $P(B)$ since this is the same for each.

Data Analysis: A Gentle Introduction for Future Data Scientists. Graham Upton and Dan Brawn, Oxford University Press.
© Graham Upton and Dan Brawn (2023). DOI: 10.1093/oso/9780192885777.003.0010

Example 10.1

As an example we return to the 2201 unfortunate individuals on the *Titanic*. A summary is given by Table 10.1.

Table 10.1 Summary of the outcomes for the individuals on the *Titanic*.

Status	Gender		Age		Outcome	
	Male	Female	Child	Adult	Died	Survived
1st class	180	145	6	319	122	203
2nd class	179	106	24	261	167	118
3rd class	510	196	79	627	528	178
Crew	862	23	0	885	673	212
Total	1731	470	109	2092	1490	711

Imagine that an unidentified adult male survivor is plucked from the water. We need to decide on his probable status.

The three background variables are gender (B_1), age (B_2), and outcome (B_3). The first class to consider (C_1) is (appropriately) '1st class'. There are 325 1st class passengers, and the probability of a randomly chosen one being male is 180/325. Similarly the probability of being an adult is 319/325, with the probability of survival being 203/325. The probability of an individual being a 1st class passenger is 325/2201. Combining this information gives:

$$\frac{180}{325} \times \frac{319}{325} \times \frac{203}{325} \times \frac{325}{2201} = 0.050.$$

The results of the corresponding calculations for the three other classes are as follows: 2nd class, 0.031; 3rd class, 0.051; crew, 0.094. The value for crew is the greatest, so we would classify the individual as being a crew member.

We can repeat the exercise supposing that the rescued individual was a female. In that case our conclusion would be that she was a Class 1 passenger. When we turn to the case where the rescued individual is a child, we meet a difficulty, since there are no crew who are children. The usual solution to this problem is to replace the 0 by a 1. With this change, for each gender, a rescued child would be classified as a 3rd class passenger.

The background information for the *Titanic* example was based on counts. Often, however, that information comes from measurements of continuous variables. In these cases, rather than calculating conditional probabilities, we can work with probability density functions (Section 4.6). Denoting the background (explanatory) variables by X_1, X_2, ..., X_m, in place of Equation (10.2) we use:

$$f(x_1) \times f(x_2) \times \cdots \times f(x_m). \qquad (10.3)$$

The most common choice for the density function is a normal distribution (Section 5.3.1) with the mean and variance chosen to match those of the available values of X. Thus usually a different normal distribution will be used for each background variable (and the calculations will be done automatically by the computer algorithm).

Example 10.2

As a simple example for the remainder of this chapter, we will use a data set published by R. A. Fisher in 1936. The data consist of measurements of the lengths and widths of the petals and sepals of three types of iris: *Iris Setosa, Iris versicolor,* and *Iris virginica.* As a first step, Figure 10.1 shows boxplots for the four measurements.

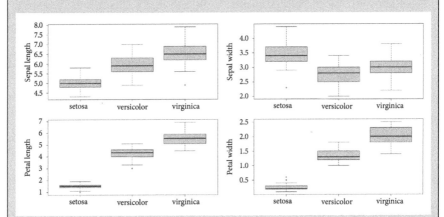

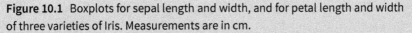

Figure 10.1 Boxplots for sepal length and width, and for petal length and width of three varieties of Iris. Measurements are in cm.

Since the petals of *Iris setosa* are much smaller than those of the other varieties, distinguishing this species is trivially easy. We will focus on the remaining two varieties.

There are 50 specimens of each variety. In order to assess the effectiveness of the various classification methods, we divide the data into two parts: a **training set** and a **test set**. We use the training set to determine the estimated values of quantities required for the classification procedure (e.g., the means and variances of the background variables). We then use the resulting classification procedure on the test set and observe the proportion of cases that are correctly assigned.

For the iris data, the first 25 cases (per species) will be used as the training set, with the second 25 forming the test set.

As the boxplots suggest, the distributions of each of the four measurements are approximately symmetric, so that the assumption that the variables are normally distributed is entirely appropriate. Calculating the value of Equation (10.3) by hand is straightforward but would be very tedious. Using an appropriate computer package the results (summarized in Table 10.2) are instant.

Table 10.2 Results of predictions made using the naive Bayes procedure.

	Actually *Iris versicolor*	**Actually** *Iris virginica*
Predicted *Iris versicolor*	24	2
Predicted *Iris virginica*	1	23

Although the four measurements are undoubtedly closely linked, the procedure, which assumes their mutual independence, has been highly effective. As a summary, Youden's index (Section 7.2.3) is $\frac{24}{25} + \frac{23}{25} - 1 = 0.88$.

10.2 Classification using logistic regression

We introduced logistic regression as a procedure for estimating an unknown probability that was dependent on one or more explanatory variables. If we are simply looking for a 'yes/no answer', rather than a numerical probability, then we can say that if $p > 0.5$ the class is 'Yes', and otherwise the class is 'No'.

Example 10.2 (cont.)

We begin by labelling the two species with the values $y = 0$ and $y = 1$. Denoting the four measurements by x_{sl}, x_{sw}, x_{pl}, and x_{pw}, we arbitrarily start with the straightforward logistic model:

$$y = \alpha + \beta_{sl}x_{sl} + \beta_{sw}x_{sw} + \beta_{pl}x_{pl} + \beta_{pw}x_{pw}.$$

Comparing the *AIC* value for this model with those for simpler models, suggests that it is the widths rather than the lengths that are most important in distinguishing the species, with petal widths being most important. Using all four measurements results in all 50 members of the test set being correctly identified. However, it will not be easy to explain the model!

10.3 Classification trees

A classification tree consists of a series of yes/no inquiries such as 'Is the colour red?', 'Is the individual an adult?' Each inquiry divides a single group of observations into two smaller subgroups. There are no parameters to be estimated (so that the method would be described as being **non-parametric**), but a necessary input is a list of the variables involved. The potentially complex rules governing the tree divisions depend on the computer algorithm used, with improvements being focused on avoiding overfitting the training set. So far as the user is concerned, the great advantage of this method is that it is simple to implement and the result is easy to understand.

Example 10.2 (cont.)

For the iris data the classification tree (Figure 10.2) resulting from the training data is remarkably simple, since it makes use of only one of the four variables. If the width of a petal is less than 1.7 cm then the diagnosis is that the petal came from *Iris versicolor*. If this is not the case then the variety is predicted to be *Iris virginica*. For the training set this classification tree is correct for 48 of the 50 cases.

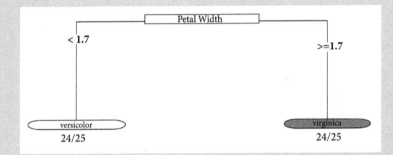

Figure 10.2 The classification tree for the iris data.

The results when the tree is applied to the test set are summarized in Table 10.3.

Table 10.3 Results of predictions made using the classification tree.

	Actually *Iris versicolor*	**Actually** *Iris virginica*
Predicted *Iris versicolor*	24	3
Predicted *Iris virginica*	1	22

The simple rule 'It is *Iris versicolor* if the petal width is less than 1.7 cm' is easily understood and implemented. It requires only one measurement and is very effective (Youden's index is $\frac{24}{25} + \frac{22}{25} - 1 = 0.84$).

Example 10.3

The classification tree for the iris data was unusually simple. For an example of a more complex classification tree we return to the baby weight data of the previous chapter. This time we divide the data into two parts, with 149 randomly chosen observations being used as the training set and the remaining 40 cases being the test set. A preliminary investigation using stepwise selection (Section 9.2.5) from a model that included all eight available variables, resulted in the choice of a model that included six variables: the mother's age and weight, her race (using the reduced two categories), whether she was a smoker, whether she suffered uterine irritability, and whether she had a history of hypertension.

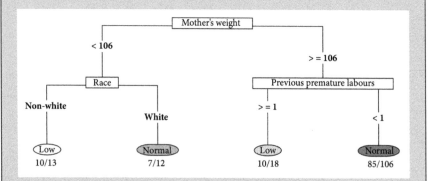
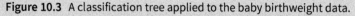

Figure 10.3 A classification tree applied to the baby birthweight data.

Figure 10.3 shows the tree obtained using the R package `rpart` applied to the training data. Notice that the tree makes use of only three of the six variables considered. The tree is read as follows:

Q1: 'Is the mother's weight less than 106 kg?'

   If the answer is 'Yes', then go to Q2.

   If the answer is 'No', then go to Q3.

Q2: 'Is the mother non-White?'

   If the answer is 'Yes', then the decision is that the birthweight will be Low (this was the case for 10 of 13 mothers in the training set).

   If the answer is 'No, then the decision is that the birthweight will be Normal (this was the case for 7 of the 12 births in the training set).

Q3: 'Has the mother previously had a premature baby?'

   If the answer is 'Yes', then the decision is that the birthweight will be Low (this was the case for 10 of the 18 mothers in the training set.)

   If the answer is 'No', then the decision is that the birthweight will be Normal (this was the case for 85 of the 106 mothers in the training set).

When applied to the 40 cases in the test set, the tree provides the correct decision in 25 cases.

10.4 The random forest classifier

The previous section focused on a single tree; we now allow the possibility of many trees. Many trees make a forest. But how can we can derive many trees from a single set of data? The answer is to apply the bootstrap (Section 5.5.1). to the training set. Each new set of data produces a new tree. The decision made is based on the aggregate information provided by the collection of trees (this is referred to as **bagging** which is short for bootstrap aggregation).

Example 10.3

The classification tree for the iris data was based solely on the width of the petals. When we use the bootstrap the resulting trees could involve other variables: Figure 10.4 shows one of the alternative trees.

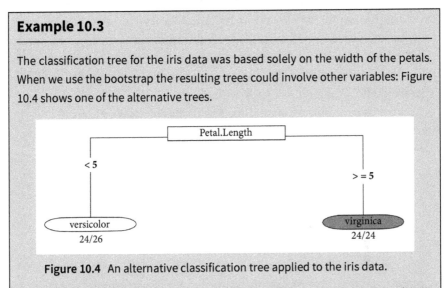

Figure 10.4 An alternative classification tree applied to the iris data.

Examination of twenty trees suggested that very few, if any, relied on sepal measurements. The effectiveness of the consensus classifications based on 500 trees are summarized in Table 10.4.

Table 10.4 Results of predictions made using the random forest.

	Actually *Iris versicolor*	**Actually** *Iris virginica*
Predicted *Iris versicolor*	23	2
Predicted *Iris virginica*	2	23

In this case the forest has failed to improve on the single tree. In more complex cases an improvement would be expected.

> It is a good idea to examine a few individual trees so as to get a feel for which variables
> are truly critical.

10.5 *k*-nearest neighbours (*k*NN)

Large data sets with information on many variables will often include missing values. A simple solution is to ignore every case for which information is missing. However, sometimes this will turn a large data set into a small data set with much useful information being unused. In such cases the solution can be to restore the data set to its full size by replacing the missing values by 'educated guesses'; this process is called **imputation** and there are several standard techniques. One simple technique is to replace a missing value by the corresponding value from the data item that it most closely resembles. That would be an example of *k*NN with $k = 1$.

As a trivial example, suppose there are just two classes (✓ and ✗) and that we have the following data:

Data Item	1	2	3	4	5	6	7	8	9	10
x	13	18	40	42	46	51	60	66	70	42
Class	✓	✓	✓	✓	✗	✗	✗	✗	✗	?

The class of the 10th data item is missing, but, by good fortune, we have another data item with exactly the same *x*-value. For that item, the class was ✓. So we use ✓ to replace the missing class. This is an example of using the *k*-nearest neighbours approach with $k = 1$.

Notice that there is no underlying model here, just an application of common sense. Since no parameters have been estimated, this is another **non-parametric** procedure.

With a lot of data it would be sensible to choose a larger value of *k*. Suppose we choose $k = 5$ instead of $k = 1$ for our current example:

Data Item	3	4	5	6	7	10
x	40	42	46	51	60	42
Class	✓	✓	✗	✗	✗	?

Items 3 to 7 have the x-values that are closest to the x-value of item 10. We see that the classes of these five are split three to two in favour of ✗, so that is the class that we infer for item 10.

In reality there will be many explanatory variables. To see how to deal with this situation we introduce a second explanatory variable, y, to our simple example:

Data Item	1	2	3	4	5	6	7	8	9	10
x	13	18	40	42	46	51	60	66	70	42
y	0	200	250	900	500	800	1000	900	700	600
Class	✓	✓	✓	✓	✗	✗	✗	✗	✗	?

We can assume that x and y are equally relevant to the determination of class, but they are measured on very different scales: the x-values range 13 from 70, whereas the y-values ranging from 0 to 1000. In order to combine the information we need to standardise (Section 9.5.1) each variable by subtracting its mean and dividing by its standard deviation.[2] We obtain x^* and y^* given by:

$$x^* = \frac{x - \text{mean}(x)}{\text{sd}(x)}, \quad \text{and} \quad y^* = \frac{y - \text{mean}(y)}{\text{sd}(y)}.$$

The results of the standardisation are as follows:

Data Item	1	2	3	4	5	6	7	8	9	10
x^*	−1.63	−1.37	−0.26	−0.16	0.05	0.30	0.75	1.06	1.26	−0.16
y^*	−1.62	−1.06	−0.92	0.88	−0.23	0.60	1.16	0.88	0.32	0.04
Class	✓	✓	✓	✓	✗	✗	✗	✗	✗	?

The nearest–neighbours in terms of their x^*-values were items 3 to 7, but these are not the items with the nearest y^*-values. The latter are 4, 5, 6, 8, and 9.

We need to combine the information from the two variables so as to get a single distance. Two methods commonly used are the so-called **Manhattan distance**, also known as the **city-block distance**. and **Euclidean distance**.

---

[2] This will be done automatically by the computer. It is shown here so as to provide a glimpse behind the scenes.

The distance between points i and j is given by:

Manhattan distance: $|x_i^* - x_j^*| + |y_i^* - y_j^*|$

Euclidean distance: $\sqrt{(x_i^* - x_j^*)^2 + (y_i^* - y_j^*)^2}$

For data items 1 and 10, these distance methods are illustrated in Figure 10.5 by the thick lines between the two points.

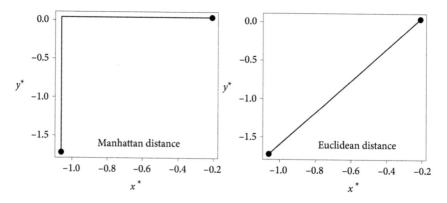

Figure 10.5 Two methods of determining distance between two points.

For our data the five nearest items are numbers 3 to 6, together with item 7 (Manhattan) or item 9 (Euclidean). Whichever distance method is used, the class assigned to item 10 would be (by a majority of 3 to 2) ✗.

Usually there will be information on more than one explanatory variable, with some variables being categorical. Suppose that, in addition to the values of x and y, we also have information on colour (red, white, or blue):

Data Item	1	2	3	4	5	6	7	8	9	10
Red	1	1	0	0	1	0	0	0	0	0
White	0	0	0	1	0	1	0	1	1	1
Blue	0	0	1	0	0	0	1	0	0	0
x	13	18	40	42	46	51	60	66	70	42
y	0	200	250	900	500	800	1000	900	700	600
Class	✓	✓	✓	✓	✗	✗	✗	✗	✗	?

Notice that, using dummy variables (Section 9.2.3), all three colour categories are included, with 1 signifying that the individual has that colour (so item 10 was white). The analysis proceeds as before. First, standardize each variable (including the colour variables). Second, calculate the distances (there will now be five distance components to add together; one each for Red, White, Blue, x, and y). Finally, identify the k = 5 nearest neighbours and determine the outcome of their 'vote'. In this case the nearest neighbours are numbers 4, 5, 6, 8, and 9. Thus the decision is again ✗; this time by a majority of 4 to 1.

Example 10.3 (cont.)

Applying kNN classification to the training data (with k = 5 and Euclidean distance), gives the results shown in Table 10.5.

Table 10.5 Results of predictions made using the five nearest neighbours.

	Actually *Iris versicolor*	**Actually** *Iris virginica*
Predicted *Iris versicolor*	23	4
Predicted *Iris virginica*	2	21

The procedure is simple, but, in this case, not quite as effective as the classification tree (Youden's index value is $\frac{23}{25} + \frac{21}{25} - 1 = 0.76$).

10.6 Support-vector machines

The training set consists of objects, each of which belongs to one of two classes. We have information on the values of m variables for each object. The aim is to use that information to decide on the class of an object, O, for which the values of the m variables are known but the class is unknown.

When m = 1 there is just a single measurement, x, for each object. A likely rule would be that if the value of x is greater than x_{crit} then O belongs to one class, and otherwise it belongs to the second class. In this case the 'support vector' is the single value x_{crit}.

Example 10.3 (cont.)

For the iris data, the classification tree suggested that petal width was the most important measurement. Using the training set suggests the choice x_{crit} = 1.65 which results in a single error for both varieties (see Figure 10.6). However, when applied to the test set, this choice is less successful, with one *Iris versicolor* specimen being classified as *Iris virginica* and with three *Iris virginica* specimens being classified as *Iris versicolor*.

In this case the 'support vector' is the single number 1.65, which determines any unknown species.

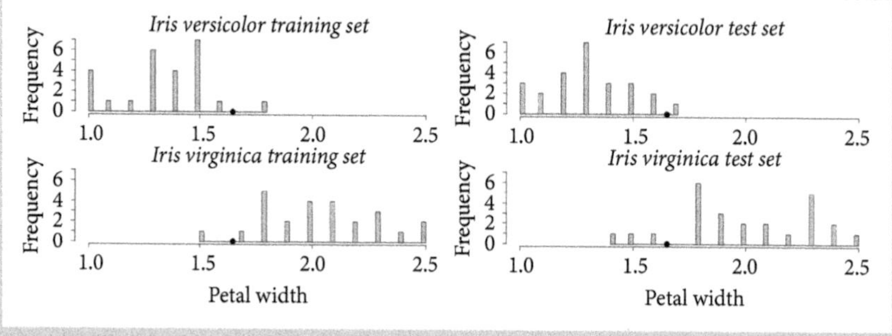

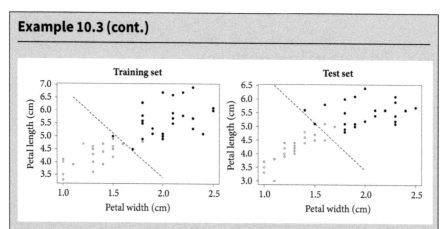

Figure 10.6   Bar charts showing the petal widths of the two iris varieties in the training and test sets. Irises with widths greater than 1.65 cm are classified as *Iris virginica*.

In the next simplest case, where $m = 2$, the separation into classes uses a dividing line rather than a dividing point, since the information is now two-dimensional.

Example 10.3 (cont.)

Figure 10.7   Scatter diagrams showing the petal lengths and widths of the two iris varieties in the training and test sets, with the solid dots corresponding to *Iris virginica*.

Since the most important measurements appear to be those of the petals rather than the sepals, we now introduce petal length as a second variable. The two varieties are quite well separated in the training set with just two misclassifications, as illustrated in Figure 10.7. However, there are three errors when the same dividing line is applied to the test set.

If there were three measurements available, the support vector would define a plane that attempted to separate the classes into distinct clouds in three-dimensional space. If $m > 3$ then the attempted division into classes would be made by a so-called hyperplane in m-dimensional space. No illustration is available!

Example 10.3 (cont.)

Using all four measurements the number of errors is reduced to two (Table 10.6):

Table 10.6   Results of predictions for the iris data using a support vector machine based on all four available measurements.

	Actually *Iris versicolor*	**Actually** *Iris virginica*
Predicted *Iris versicolor*	24	1
Predicted *Iris virginica*	1	24

The separation is very effective with a Youden's index value is $\frac{24}{25} + \frac{24}{25} - 1 = 0.92$.

10.7 Ensemble approaches

Do you wonder what the weather will be like later this week? If so, then you probably look at a weather forecast. But do you believe the forecast? Possibly not. Perhaps you use several weather apps and then form your own opinion. If so, then you have been using an ensemble approach: you have been, in a sense, averaging the outcomes of different models.

Exactly the same approach can be used in the context of classification. If we use several methods and they all agree, then we will be rather confident with their joint decision. If they disagree, then we should be doubtful about whichever classification we choose.

Example 10.3 (cont.)

For the iris data we have used six distinct classification methods. If the predictions split 3–3 we will call the result undecided, otherwise we will accept the majority verdict. The results are summarised in Table 10.7.

The ensemble approach cannot improve on the perfect diagnosis produced by the logistic regression approach for this data set. However, we must remember that, in practice we would not know the true classes of the objects being classified, so that it is reassuring that 47 of the 49 decisions made by the ensemble approach were correct.

Table 10.7 Results of predictions made using the six classification procedures.

| | \multicolumn{5}{c}{Decision split} | |
	6–0	5–1	4–2	3–3	2–4	Total
Correct	42	2	3			47
Undecided				1		1
Incorrect					2	2

The reader should be aware that no single example can be regarded as providing a ranking of the effectiveness of these methods. In practice it would be wise to experiment by using other values of k for the nearest neighbour approach, alternative logistic models, or alternative variable selections. We can also combine variables together (see below).

10.8 Combining variables

In the examples, we used the four variables as they were measured. However, it may be useful to use functions that combine two or more variables.

Example 10.3 (cont.)

If we multiply length by width then we get a measure of area. If we divide length by width then we get a measure of shape. If we divide sepal length by petal length, then we may get information on the appearance of the flower. Including these three derived variables results in perfect predictions from every method except the kNN method. There, with $k = 5$, the results improve by reducing the number of incorrect classifications from six to two.

11
Last words

The intention of this book has been to introduce the reader to methods for data analysis without dwelling on the underlying mathematics. We have introduced a great many methods that may help with the data scientist's task of forming testable explanations of the data. With any new data set it is always difficult to know where to start. Here is some advice:

1. Go slowly!
2. Learn where the data have come from (sample, questionnaire, …) and how they were obtained (by a machine, by interview, by a person taking measurements, …)
3. If feasible, consider taking some more data yourself.
4. Remember that most data sets are dirty: they are likely to have errors (e.g., some measurements in metres, others in kilometres)
5. Plot the data for each variable (use a bar chart, histogram, box plot….) to look for that dirty data.
6. Perhaps calculate correlations between pairs of numerical variables, and plot scatter diagrams in order to get a feel for the relationships between the measured variables.
7. Use box plots to compare numerical values when one of the pair of variables is categorical.

Remember that curiosities in the data may have an explanation that requires none of the methods of the last few chapters.

Example 11.1

Here is a tale, possibly apocryphal, about some time-series data (measurements of some quantity taken at regular intervals in time). The industry concerned had hired a consultant to analyse the data because there was concern that the early values were on average larger than the later values. The consultant asked to be shown how the data were obtained. The answer was that the daily values were taken by an observer noting the reading of a pointer on a dial placed high on the wall. Further questioning

---

Data Analysis: A Gentle Introduction for Future Data Scientists. Graham Upton and Dan Brawn, Oxford University Press.
© Graham Upton and Dan Brawn (2023). DOI: 10.1093/oso/9780192885777.003.0011

revealed that there had been a change in the staff member responsible for taking the reading: a short person had been replaced by a tall person. The apparent change in the levels recorded was simply due to the change in the angle at which the observer read the dial.

Example 11.2

Earlier, in Example 8.2, we examined the relation between the eruption lengths and the times between eruptions for the Old Faithful geyser in the Yellowstone National Park. The first few records of the eruption lengths are apparently recorded in minutes correct to three decimal places:

3.600   1.800   3.333   2.283   4.533   2.883.

What do you notice?

> *Always be suspicious of your data! Study it closely.*

In this case it is evident that the data were not originally recorded in thousandths of a minute! They were recorded in minutes and seconds.

> *It is often worth recreating the original data, since this may highlight shortcomings in the recording procedure.*

We might wonder whether there is evidence of unconscious 'round-number bias' in the number of seconds recorded. In order to select the seconds component we subtract from each data item the number of minutes and multiply the remainder by 60:

36   48   19.98   16.98   31.98,   52.98.

It is easy to see that the computer has introduced round-off errors: for example, 19.98 should obviously be 20.

Since there were 272 eruption durations recorded, and (after rounding) there are 60 possible values that can be recorded for the number of seconds, we might expect that each value would occur on about 272/60 occasions. Figure 11.1 presents a bar chart of the frequencies of the numbers of seconds recorded. The white line indicates the average frequency.

There is definitely evidence of 'round-number' bias, since all of 0, 10, 20, 30, and 50 have larger than average frequencies. This is also true, to some extent, for 5, 15, 25, 45, and 55.

This is another example where the recorded data are not entirely accurate.

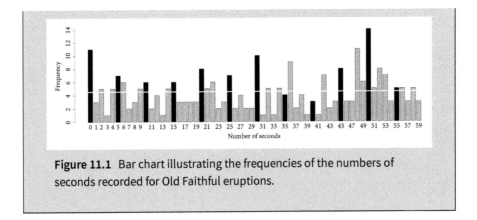

Figure 11.1 Bar chart illustrating the frequencies of the numbers of seconds recorded for Old Faithful eruptions.

Example 11.3

For each of 25 quadrats (square survey regions) on Dartmoor, students were asked to report the percentage of a quadrat occupied by various species of plant. For Heath Bedstraw (*Galium saxatile*) their reported cover values were:

0, 0, 10, 0, 1, 5, 35, 40, 5, 10, 0, 2, 10, 0, 20, 0, 5, 10, 25, 0, 0, 0, 50, 25, 0

There are only ten different values being reported:

0, 1, 2, 5, 10, 20, 25, 35, 40, 50

This is all quite natural. If a student reported, after a visual inspection, that there was 23% cover in a quadrat, then there would be hoots of laughter!

Although the inaccuracies uncovered in these last two examples are unlikely to affect any conclusions, it is always wise to bear in mind any data shortcomings ... and it shows that the data scientist has been looking at the data!

- *When there is a large amount of data, it is a good idea to look first at a more manageable small section, hoping that this is representative.*
- *Expect to need to try several different approaches before you arrive at something that appears to make sense of the data.*
- *You should always expect a data set to contain the unexpected: missing information, implausible values, and the like.*

continued

- *If the data have been taken over a period of years, or from several different regions, then you should look for inconsistencies such as measurements in different units.*
- *The key to good data analysis is to have no preconceptions and to be prepared for a good deal of trial and error.*
- *The data analyst often needs to try many diagrams before finding the one that makes the most useful display of the data.*

And finally just two things remain: to hope that you have found our tiny tome helpful, and to wish you good fortune in your career as a data scientist.

Further reading

- Agresti, A. (2018). *An Introduction to Categorical Data Analysis* (3rd edn, Chichester: Wiley).
- Agresti, A. (2022). *Foundations of Statistics for Data Scientists, with R and Python* (CRC Press).
- Bruce, P., Bruce, A., and Gedeck, P. (2020). *Practical Statistics for Data Scientists* (2nd edn, O'Reilly Media Inc.).
- Crawley, M. J. (2014). *Statistics: An Introduction using R* (2nd edn, Wiley).
- Grus, J. (2019). *Data Science from Scratch: First Principles with Python* (O'Reilly Media Inc.).
- James, G., Witten, D., Hastie, T., and Tibshirani, R (2021). *An Introduction to Statistical Learning with Applications in R.* (2nd edn, Springer).
- Rowntree, D. (2018). *Statistics without Tears: An Introduction for Non-mathematicians* (Penguin)
- Spiegelhalter, D. (2020). *The Art of Statistics: Learning from Data* (Penguin).

Index

2 x 2 table, 67

Addition rule, 24
AIC, 99
Akaike information criterion, 99
Average, 10

Bagging, 127
Bayesian information criterion, 99
BIC, 99
Bimodal, 9
Bootstrap, 50
Box-whisker diagram, 13
Boxplot, 13

Central limit theorem, 45
Chi-squared distribution, 56
City-block distance, 129
Classification, 121
Classification tree, 125
Cluster sampling, 5
Coefficient of variation, 52
Collinearity, 101
Conditional probability, 25
Confidence interval, 43
    for regression line, 88
Contingency table, 68
Continuous data, 14
Correlation, 79
Cross-validation, 110
    k-fold, 110
    leave-one-out, 112
Cumulative distribution function, 40
Cumulative frequency polygon, 16

Data
    categorical, 1
    continuous, 2, 14
    discrete, 2
    multivariate, 11
    qualitative, 1
Deviance, 97

Deviation
    standard, 19
Distance
    city-block, 129
    Euclidean, 129
    Manhattan, 129
Distribution
    t-, 62
    chi-squared, 56
    exponential, 54
    extreme value, 55
    Gaussian, 44
    normal, 44
    Poisson, 53
    rectangular, 41
    standard normal, 45
    uniform, 41
    Weibull, 55
Dummy variable, 102

Ensemble, 135
Estimate
    point, 43
Estimator, 43
Euclidean distance, 129
Event, 22
    mutually exclusive, 24
Exponential distribution, 54
Extreme value distribution, 55

False discovery rate, 72
False negatives, 71
False positives, 71
Feature selection, 97
Function
    cumulative distribution, 40
    probability density, 39

Gaussian distribution, 44
Generalized linear model, 115
GLM, 115

Histogram, 14
Hold-out set, 110
Hypothesis
  null, 60

Imputation, 128
Independence, 67
Interaction, 107
Intercept, 84
Interquartile range, 12
Intersection, 22
Interval
  prediction, 88

Jitter, 94

k-fold cross-validation, 110
k-nearest neighbours, 128
Kolmogorov–Smirnov test, 65
KS test, 65

Least squares, 84
Leave-one-out cross-validation, 112
Likelihood, 43
Logistic regression, 116
Logistic regression model, 117
Logit, 116
Loglinear model, 118
LOOCV, 112

Manhattan distance, 129
Maximum likelihood, 43
Maximum likelihood estimation, 43
Mean, 10
  trimmed, 11
  Winsorized, 11
Mean squared error, 110, 113
Median, 12
Mle, 43
Modal class, 9
Mode, 9
Model
  general linear, 115
  logistic regression, 117
  loglinear, 118
  saturated, 97
MSE, 110, 113
Multimodal, 9
Multiple regression, 99

Multiplication rule, 24
Multivariate data, 11
Mutually exclusive events, 24

Naive Bayes classification, 121
Negative predictive value, 71
Nominal variable, 2
Non-parametric, 125, 128
Normal distribution, 44
Null hypothesis, 60

Observation, 3
Ockham's razor, 98
Odds, 70
Odds ratio, 70
Ogive, 16
Ordinal variable, 2
Out-of-sample validation, 110
Outlier, 11, 91
Overfitting, 97
Oversampling, 7

Pearson residual, 73
Percentile, 13
Permutation test, 85
Plot
  q-q, 63
  quantile-quantile, 63
Poisson distribution, 53
Poisson process, 52
Positive predictive value, 71
Prediction interval, 88
Principle of parsimony, 98
Probability, 21
  conditional, 25
Probability density function, 39

Q-Q plot, 63
Quantile, 12
Quantile-quantle plot, 63

Random forest, 127
Random variable, 3
Range, 12
  interquartile, 12
Rectangular distribution, 41
Regression
  logistic, 116
  multiple, 99
  ridge, 114

Regression line
  confidence interval, 88
Relative risk, 70
Resampling, 51
Residual, 73, 85, 109
  Pearson, 73
  standardised, 74
Ridge regression, 114
Rounding error, 41

Sample
  simple random, 5
Sample space, 22
Sampling
  cluster, 5
  stratified, 6
  systematic, 7
Sampling frame, 5
Saturated model, 97
Scatter diagram, 78
Schwarz criterion, 99
Sensitivity, 71
Shrinkage, 113
Simple random sample, 5
Skewed
  negatively, 20
  positively, 20
Slope, 84
Specificity, 71
Standard deviation, 19
Standard normal distribution, 45
Standardization, 114
Standardized residual, 74
Step diagram, 17
Stepwise selection, 107
Stratified sampling, 6
Support-vector machine, 131
Systematic sampling, 7

t-distribution, 62
t-test, 62
Test
  t-, 62
  Kolmogorov–Smirnov, 65
  KS, 65
  permutation, 83
Test set, 123
Theorem
  central limit, 45
  total probability, 28
Total probability theorem, 28
Training set, 110, 123
Transformations, 89
Trimmed mean, 11
True negatives, 71
True positives, 71
Tuning constant, 114

Unbiased, 43
Uniform distribution, 41
Union, 22

Validation set, 110
Variable
  dependent, 85
  dummy, 102
  explanatory, 85
  nominal, 2
  ordinal, 2
  random, 3
  response, 85
Venn diagram, 22

Weibull distribution, 55
Winsorized mean, 11

Youden's index, 71